SOURCE*FORGED ARMOR

ARMOR

Paul J. Bartusiak

This is a work of fiction. All characters, companies, organizations, agencies, and situations in this novel are either fictitious and the product of the author's imagination or, if real, used fictitiously without any intent to describe actual conduct or situations. Any resemblance to real persons, living or dead, is purely coincidental.

for Alison

CONTENTS

Crowdsource, v, - To engage the services of the general public, or a subset thereof, for the performance of a task or accomplishment of a goal, often for the creation of a work-product deliverable.

TACTICAL TECHNOLOGY OFFICE SOLICITATIONS
[Excerpt: Posted September 9, 2010]

DARPA-ZXX-10-09 MAV: <u>Tactical Technology Office (TTO) Broad Agency Announcement (BAA) for Marine Amphibious Vehicle (MAV)</u>;

The Tactical Technology Office (TTO) of the Defense Advanced Research Projects Agency (DARPA) is soliciting innovative proposals under this Broad Agency Announcement (BAA) for the performance of research and design of a Marine Amphibious Vehicle (MAV) that represents revolutionary improvements to the efficiency and effectiveness of landing surface assault elements to inland objectives during amphibious operations. <u>Response Deadline 09/30/2012</u>

IP Newswire
[Excerpt: October 11, 2010]

"A branch of the Pentagon known as the Defense Advanced Research Projects Agency (DARPA), formed over fifty years ago with a mission to maintain the technological superiority of the U.S. Military by sponsoring revolutionary, high-impact Research and Development, announced the official launch of its first ever large-scale crowdsourcing design project, codenamed ZXX-10-09 MAV. The project centers on the design of a next-generation vehicle for the transportation of the Marines and their equipment inland from sea waters. The vehicle is called a Marine Amphibious Vehicle, or MAV, and past efforts by the Defense Department to procure a new design for the vehicle through traditional procurement methodologies has been met with unsatisfactory results, including projected costs that would far exceed the budgetary limits for such a…."

1

A Clandestine Submission

NOVEMBER 28, 2011

Standing at the window of his eleventh floor office, the Professor watched as a thick snowfall covered the city. It was early evening, and the sky was a toneless grey.

How peaceful the view would have appeared to the casual observer—almost inspiring: people busily making their way through the city, bundled in their heavy, winter clothes as they fought against the weather, cars buzzing in every direction with headlights glaring, and streetlamps emitting a soft glow as flakes fell through their illumination, all of this against the backdrop of the grand, historic buildings of St. Petersburg, Russia.

Not for him. He had seen enough, suffered enough, and could no longer appreciate the beauty of it. He was turning his back on all of it, or at least, everything that was left of it, his country.

He had been standing there, gazing out of the window for a long time in deep reflection, contemplating what was left of his life, and what he had lost, and the country that, to him, no longer existed. He was overwrought with the inner pain that he carried. His spirit was low, and yet...there was still *something* left to sustain his purpose.

Behind him in that office were stacks of documents and journals piled everywhere. Technical books lined the shelves in testament to his intellect. As an esteemed Professor of Engineering, his office was so crowded with books, papers, samples, and whatnot, that there was barely enough room for the guest chair in front of his desk, and if someone were to sit in that chair, he or she would have to lean sideways in order to see, through the piles of books and journals, the Professor himself.

It was Friday, the end of the academic week, and late enough in the evening for most of the students and faculty to be gone from the building. Thus, all that could be heard were the sounds of life outside as it permeated the building walls and traveled into his office. Thoughts preoccupied him so much that he was oblivious to the noise.

Amongst the overabundance of materials in that office, one item held his awareness; it was resting indistinctly on an old, wooden credenza at the other side of the room. He turned to face it, and the power and gravity of it weighed on him. That item, which had hidden within it the key or pathway to something extraordinary, had been like a pulsating power source, generating waves of energy that constantly assaulted him and reminded him of its presence.

He took a deep breath, and then a step toward it, and for the slightest moment he wondered whether he might have a change of heart. It was a fleeting, ridiculous notion which, if left to develop to its fullest, would nonetheless have withered in the measure against all that ultimately weighed against it. The sentiment did not, however, have an opportunity to become manifest, because just at that moment someone appeared at the door—a man holding a large, empty envelope under his arm.

The winter coat on him was wet from snow that had already melted from the building's heat, and his black hair was matted down, the culprit, his moist winter hat, held in both of his hands in a nervous state of tension. The man raised his hand to knock on the doorframe but stopped when he saw the Professor standing in the middle of the room, and he could not help but notice the distressed look on his mentor's face, with the grey, wiry hair combed back on the Professor's head, more disheveled than ever, and wrinkles on his face that took the form of deep crevices. It was a bittersweet moment for both of them; they were setting upon a course that could change lives forever, under circumstances for which they wished were not so. The man, who was essentially the Professor's protégé, had been walking outside, making his way through the bustle of the city toward the University, with the ultimate purpose of his visit blotted out from his consciousness. The reality of the impending event, however, and the gravity of its ramifications, which the Professor had just then been fully considering, hit the man at the moment the two gazed at each other.

"Will it be as we discussed?" the Professor asked in a grave tone (and in their native, Russian tongue).

"Yes, just as you instructed. Tonight at the Marinsky Theater," the man replied as he entered the office and walked toward the Professor to give him the envelope.

"Good."

They again looked at each other in silence, with thoughts passing between them: memories of times past, and concern for an unknown future. Each considered extending a hug to the other, but both resisted the urge in an attempt to display courage and command of their emotions.

The Professor reached for and received the thin, cardboard envelope, walked to the credenza, and gently tossed the envelope on top of the *item* — the one that had been resting there with all of its power and force. After a pause, he slowly extended his arms, put his hands on the sides of the credenza, and allowed his head to droop forward, as if by the weight of the circumstances. The office grew darker by the minute as the small amount of natural light that came through the window diminished with the approaching night. Without any light in the office turned on, each of them could only see the faint outline of the other from the distance that separated them.

Observing the body language of the Professor—the slow movement, body hunched over, and head slumped downward—created a solemn mood in the protégé, so much so that it was as though the temporary doubt that had momentarily crept into the Professor's mind had transferred itself to the other man, compelling him to say, "I'm sorry to ask, Professor. We've gone over it so many times. Are you... are you still sure? I mean, it's not too late for us to stop."

He somewhat regretted verbalizing the question as soon as he had asked it, given everything that had been done up until that point, and the critical stage that they were at in the process, but he wanted to know the Professor's thoughts. He was compelled to ask because soon it would be too late for them to change their minds.

With his head still bent forward, the Professor raised one of his hands to his forehead and massaged it slowly. "I'm only sure of one thing..." he started to say, but then stopped. The office was quieter than before; the thick snowfall on the ground by that time muffled the noises that would have otherwise been more prominent from outside. "It's not my...our...country anymore."

It was surprising to the Professor how difficult it was for him to say those words, because they were expressed between them so many times before. He turned from the credenza to face the young man who was so many years his junior, and with an attempt to appear more in control of himself, he said, "And yes, it's too late to stop. Matters have been put in place, and risks, large risks, have already been taken." The Professor walked over to his desk and reached to turn on a small desk lamp, and Dmitri was slightly startled by the sudden light. He saw the Professor curl one of his index fingers and rub the side of it against the corners of each of his eyes, wiping away trace amounts of moisture.

"Others are depending upon us and have placed themselves at risk," the Professor continued. He cleared his throat, and in an effort to transition the conversation to the matter directly at hand, asked, "Are you sure that you're done queuing up the information for staged release?"

Dmitri noticed the Professor's more formal, businesslike demeanor and said, "Yes, yes. Of course. It's all established. I've stepped through and verified the whole process. Our software coder did an excellent job." Then, with slight hesitation, because he knew that it was a subject that he had broached with the Professor so many times before, he said, "I still don't know the names of the other contributors."

"I know. Soon," the Professor responded.

Soon it would be a matter of luck, the Professor thought to himself...being selected in a sweepstakes, as it were...being noticed in a crowd of performers...standing out. It was the only way they could see doing it—one chance in a million, perhaps. He wondered whether their voice would be heard. Whether what they were going to send would be discovered by its intended audience.

He picked up the item from the credenza—the one that had been resting there all day and commanding so much of his attention and concern—placed it into the courier envelope that Dmitri had given him, and sealed it.

Now he needed some courage.

He went over to a cabinet and opened it to reveal a small sound system. Dmitri watched as the Professor bent down, raised his eyeglasses onto his forehead so that he could read the display on the unit, and when he found the selection that he was looking for, hit the play button. It was Prokofiev's *Dance of the Knights*. Slowly he turned the mechanical dial to raise the volume ever so slightly, and for a moment the Professor stood with one of his hands resting on a top surface of the cabinet and listened, his mind trailing off to somewhere in the distance. Tears began to well up in his eyes again, but he caught himself and wiped them away with his shirt sleeve. He was agitated with himself over how emotional he was becoming that evening. He bent down again, reached into the cabinet, and took out a bottle of vodka and two glasses.

As he poured the drinks, he looked up at his protégé with a wry smile, and in an attempt to lighten the mood he said, "I can tell by the look on your face, Dmitri, that you're not pleased with my music selection. Are you still confused by your fondness for Italian composers over those of your own country?"

The Professor offered one of the filled glasses to his colleague, who in turn came closer to accept it. The subject of classical music was a friendly, ongoing debate between them that they perpetuated in jest.

They raised their glasses in a toast, and after they each took a drink, Dmitri responded, "No, Professor, not confused; most assured."

The Professor was glad that Dmitri had accepted his invitation to joust, and he said, "And who might you have chosen instead for this momentous occasion, if I may be so bold as to ask?"

Dmitri held up his glass to look at it, and it was still dark in the room so that he could barely discern its contents. He walked closer to the desk lamp, held his glass up again to look at it, and then turned to face the Professor with a serious look. "For once, I am at a loss as to what would be appropriate for the occasion." The smile on the Professor's face disappeared, and seeing that, Dmitri regretted saying what he had; he became determined and looked down at the ground for a moment in thought. The Professor watched, and soon Dmitri looked up and said, "I believe that your choice was too obvious. No, for tonight I would have something else in mind: Bartok's *Two Portraits, opus five*."

"Hmm...Hungarian?" said the Professor with a contemplative expression. It was an unexpected choice; perhaps a concerted attempt by Dmitri to draw attention to the extraordinary nature of the present situation. "I see," the Professor said as he stroked his long, grey beard and took a deep breath. "A compromise of sorts, eh? Well, we won't comment further beyond that."

Dmitri held his glass up to the Professor in homage. The Professor returned the gesture and took a last gulp to finish his drink. Then he put his empty glass down, walked over to grab a black tuxedo that was hanging on the inside of his office door, and began to change into it. His old, heavyset body made it a slow process.

Dmitri poured himself another drink in the meantime and went over to the window to look out of it while the Professor changed. He was tall, and had deep black hair that was parted in the middle of his head, with long bangs that were pulled to the sides behind each ear; the hair was shoulder length. His nose was straight, large, and puffy. Being in his mid-thirties, much life was still ahead of him, and he knew it. It was something that he pondered on a daily basis as he prepared for what they were about to do. The pedestrians and cars passed by below him, marching down the sidewalks and speeding down the streets. They looked small to him as he gazed down at it all from so high up in the building.

When the Professor was fully dressed, Dmitri helped him carefully place the sealed envelope into a specially crafted pocket of the inside back liner of the Professor's winter coat, and when the Professor was finished putting on the coat, he stood in front of Dmitri. The music was still playing quietly in the background. They looked at each other, wanting to say something further, but did not, and instead they relented and hugged each other. The Professor again had to fight back tears. It was an emotional moment for him, and he was already in an emotional state as it was.

He broke from his protégé abruptly. "It's time for *Boris Godunov*," he said in a gravelly voice and with an attempt to inject a tone of conviction.

* * *

When the Professor stepped outside of the building of the Saint Petersburg National Research University of Information Technologies, Mechanics and Optics, the cold hit him immediately. He decided that even though it was still snowing

7

heavily outside, he wanted to walk for a while on his way to the opera in order to take in the moment one last time before all might be changed forever. He buttoned the high collar of his heavy coat and pulled the flaps of his mink hat down over his ears. The blistery cold air, and the sensation of the crunch of the snow under his feet, temporarily lifted his spirits and invigorated him.

He trudged slowly and carefully along the sidewalk as he absorbed the life going on around him. At his advanced age it was difficult for him to move through the un-shoveled and heavily trampled snowy sidewalk, but for the first time in a long time, he enjoyed the effort, rubbing shoulders with other pedestrians who were out and about and moving past him.

He was able to walk for some distance before he finally saw something that brought his mood back to reality: it was a homeless man, lying on a piece of cardboard on the ground, half frozen in the snow. His heart sank, and he tried to quicken his pace and block out the image. The snow began to fall more rapidly, and as a further attempt to distract himself from the sight of the homeless man, he turned his attention to the lighted buildings reaching up into the sky, and he felt inspired by them; he watched the thick snowflakes falling through the air for a moment.

His pace was slowing; between his old age, the snow, and the bitter cold, his body could not move as his mind wanted it to. He wanted to press on just a bit further, however, and so he pushed himself.

Looking straight ahead, he saw something off in the distance; it was resting on a bench under the next streetlamp. As he approached it, he realized that it was a man sitting on top of the bench's backrest, wavering back and forth. The Professor sensed that something was wrong, and when his walk had taken him directly in front of the man, who was by then fully illuminated from the streetlight above, he looked closely and saw that the man's eyes were rolling around in the back of his head. He was intoxicated, stoned, or something, the Professor could not tell. The sight of the man made the Professor sad and again brought him back to reality, and mentally, as well as physically, it was enough. He was disheartened and could walk no more, and he hailed a cab for the rest of his journey to the Marinsky Theater.

The cab driver tried to converse with him, but the Professor was in no mood. Instead, he leaned his head against the window and stared out of it into the night in contemplation of how much more his country could have become, versus what it actually was. The sullen mood caused his thoughts to drift further from the present, until eventually that one memory surfaced that he had so often fought to suppress. It was the image of his deceased wife: how, when she was dying, she lay on her back, coughing and writhing in pain as the disease consumed her. It unnerved him, and it was all he could do to force the thought out of his head. He must remain strong and focused, he told himself.

* * *

When the taxi arrived at its destination, he entered the theater amid the stream of other patrons and went straight for the coat-check. As he stood in line, he slowly looked around in all directions, nervous to see if anyone might be watching him. When it was his turn, he stepped up to the counter and looked into the eyes of the woman that was working at the booth, as if he was asking her with his eyes something that he could not say out loud—*are you the one?* —but she barely returned his stare before she took his coat and handed him his ticket stub in a matter-of-fact manner.

One of his old friends who held a season ticket for the seat next to the Professor's for many years was already seated up in the balcony. The man looked up at the Professor with a serious look on his face and nodded to him. Ever so slightly, the Professor returned the nod before extending his hand and more openly greeting the man. The Professor knew that he must act natural— like nothing was amiss.

The lights would soon dim, and as a final check, the Professor looked through his opera glasses to observe various points around the theater to see if anyone might be watching him. There appeared to be no one, and so he took the opportunity to look around and observe the splendor of the Marinsky Theater for what would probably be his last time.

The building never ceased to amaze him, no matter how many times he experienced it. He was seated in the Dress Circle tier, only three boxes away from the Emperor's Box at the center of the theater. The light blue draping, curtains, and ceiling, highlighted by the beautifully maintained gilt, set a tone that still inspired him—if not for Russia, then for something. He contemplated with wonder the angelic fresco painted on the light-blue backdrop of the ceiling, illuminated by the grand chandelier in the center.

Then darkness came, and the applause started as the curtain began to rise. One chance in a million: that was all he figured it was. All of that hard work, planning, secrecy, and risk, for one, slight, chance. He turned to look at his colleague through the darkness, who had also turned to face him, and the man had an intense look on his face; they knew that it was very likely the last time that they would ever see each other.

Back at the coat-check area, a hand grabbed the Professor's coat and replaced it with another of similar appearance. The man who had switched them put the Professor's coat on and walked out of the theater without anyone noticing the difference.

2

Assignment for a Ghost

DECEMBER 2012

John Angstrom sat in his office in Arlington, Virginia, waiting for it to be nine o'clock in the morning. It was not his custom to be in so early; ever since he joined the TTO, it was rare for him to be in before nine-thirty. He was respectful of Seavers, though, and if Seavers wanted him in that early, Angstrom would be there.

"Have a seat, John," Seavers said as he stood behind his desk and motioned to a chair while carefully observing Angstrom.

Angstrom sat with his legs crossed and his arms resting on the chair's armrests, appearing comfortable and confident, and looked at Seavers with a blank expression on his face. At six-feet, four inches tall and a firm build, Angstrom made for quite a presence, even sitting down.

Seavers sat down as well, and nothing further was immediately said between them. There was no small talk or pleasantries exchanged, and Angstrom exerted no effort to demonstrate cordiality. Seavers considered the man that sat before him. It was typical Angstrom, he thought; no effort to lighten the mood, few words spoken, if any, and just sitting there, stoic and immobile— almost a shell of a person; a ghost.

Seavers leaned forward, picked up a document from his desk, put on his black-framed glasses, and studied the document for what seemed to him like the hundredth time. There was a multitude of facts about Angstrom on it: date and place of birth, height, weight, education, and so forth. Yet the most important information was missing, including what he had been doing for the last twenty years in the service of the United States Government. There were blank spaces all over the document where there should have been data.

It was two years ago when Angstrom was removed from his old position and put under Seavers, yet Seavers still felt like he did not know the first thing about the man. He turned his attention from the document and back to Angstrom.

"So, John, you've been in my Office for a couple of years now." The statement was delivered with the intention of serving as an invitation to Angstrom—an invitation to acknowledge the innocent declaration and follow it up with some type of response. It was a softball lodged from manager to subordinate to lighten the mood and facilitate the opening of a dialogue. Angstrom's only reaction, however, was a slight nod of the head.

Seavers immediately got the impression that if he himself did not speak, they would be left sitting there staring at each other like a couple of statues. He put the document down and picked up another one. It was a form that Angstrom had filled out prior to their meeting—a document an employee was supposed to complete in preparation for an annual performance review. It had many pre-printed questions on it, asking the employee's opinion about job performance, goals for the future, and the like, all for the purpose of facilitating a performance dialogue, which, in Angstrom's case, was now taking place.

"In your latest performance review document," Seavers continued, "you indicated that you wanted to be put on the management track." It was another factual statement framed with the intention of eliciting a response, and Seavers again looked up at Angstrom. This time, however, there was not even an acknowledgement by him. For most people, Seavers' statement would have prompted some kind of a response; thus, Seavers realized that he was witnessing first-hand what he had heard from his other direct reports so many times before as to what Angstrom was like.

"You wrote that you wanted to manage other people at the Deputy Director level," Seavers continued.

Still no response; no gesture or reaction whatsoever; just an attentive look.

Seavers thought back to when he originally read Angstrom's request on the document—to be groomed for management—and how he thought at the time that it must have been a joke; that Angstrom was finally showing a sense of humor. There could have been no other explanation, he reasoned. At present, sitting in his office and looking at the performance review form, he wanted to laugh again at the idea, but resisted the urge after not seeing the slightest hint of humor evident from Angstrom. He tossed the form back onto his desk, took his glasses off and held them in his hand, and leaned back in his chair. He decided that if Angstrom was not going to do any of the talking, then he was going to have to carry the conversation himself and be frank with the man.

"Listen, John, we've never really had a long talk before. I was instructed not to ask too many questions when I was originally informed that you were joining my Office, and we've said very little to each other ever since that time. In fact, I've purposefully kept at a distance out of respect, based upon the vague *signals* that were provided to me from the Director. Last year, I even skipped your performance review. But it looks like you're going to be with us for a while now, and I think it's time that we finally had a good conversation with each other."

Seavers' manner in which he said what he did triggered something in Angstrom; the directness of it affected him, and Seavers noticed a small spark in Angstrom's eyes.

"Sure, Bill. I appreciate the opportunity."

Even those few words felt odd for Angstrom to say. It took great effort, and he had to force himself. He was a man of few words as it was, and it was even harder for him given the circumstances of his present life. He had been relegated to the proverbial desk job; someplace *safe*; a landing spot for damaged goods. It felt to him like he was in detoxification, because he could not have been any further from what he had been doing as a field operative for over twenty years. He was pushing papers now, and going through the motions—watching the clock tick by, day in and day out. After two years of it, however, he finally resolved to himself that his old life was over, and that his job in the TTO was his new reality. He needed to face that fact head-on and force himself out of the shell under which he had been living. It was time for him to regroup and forge some semblance of a life—that was why he included what he did on the performance review document. Moreover, after working in the TTO for two years, he could tell that Seavers was a good man, someone who cared about the people that worked for him. So Angstrom was ready to take that first step with him.

Seavers, as of late, had been dropping by Angstrom's office periodically with the purpose of trying to stimulate conversation between the two of them, and the effort seemed to have worked; it moved Angstrom out of his detached state of existence. Angstrom could tell that Seavers was making an effort to understand him.

Seavers was in his early sixties, at an age when many people would have been coasting in their government jobs and thinking about the next phases of their lives. That was not the case for Seavers. He enjoyed his position at the TTO and was still very interested in all of the research and development projects for which his Office was responsible. His organization was at the forefront of advanced military technology, and he held great respect for the responsibility. He was dedicated.

"From the way that our Office was strongly urged to *adopt* you, and the ensuing scant amount of information that accompanied your landing here, I can only surmise that you must have been one hell of a field operative," Seavers said. He was never actually told that Angstrom was a field operative, but he had guessed as much, and he watched Angstrom closely to see if he could detect at least some non-verbal reaction that might, inherently, reveal the answer to his conjecture.

But Angstrom was too well trained, too experienced, for a statement like that to influence him, and he remained unmoved. His training instinctively kicked in; it was automatic. His past was beyond mention, and he was not allowed to even confirm such speculation. As a result, he turned his head away and looked at some remote aspect in the room. He was not ignoring Seavers; he was considering a way to continue the dialogue, and what he *could* talk about.

Even though Angstrom's reaction was not an involuntary response to Seavers' veiled probing, Seavers nonetheless misinterpreted it as such and believed that his suspicion had been confirmed. He was therefore not offended by what otherwise may have been considered a rebuff when Angstrom turned his head away. He was being patient with the man, knowing in advance the deliberate manner in which Angstrom conducted himself and the sparse style of speaking which was his nature. Seavers had been working at DARPA for a long time and had met a lot of different types in his day. He was not easily offended, and in actuality, at that point in his career, he was experienced enough to discern that Angstrom was not his normal self, that something was not right with him, and he was considerate of Angstrom's apparent condition.

"John, I'm going to be frank with you. I don't know what happened to you, or why you're assigned to me…and I'm hesitant to say this, but all I see in you is a ghost, sometimes here, sometimes not. Engaged just enough to complete what is asked of you, and no more. Furthermore, and I'm sure you must realize this, your overly reserved manner either puts your colleagues off, or just plain creeps them out."

Seavers saw that Angstrom was listening carefully and did not seem to be angered by what he said, so he continued, "I can't put you on the Associate Director-level track within DARPA. Surely you must know that. I mean…it's not like your fresh for the party anyway, you know what I mean? You're closer to fifty than you are to forty, and I'm guessing that you worked alone for most of your career rather than supervised others." Seavers knew that he was not supposed to allude to one's age in the context of the present conversation, but he felt like he could speak freely with Angstrom, and his instinct was usually right.

Angstrom processed what was just said, and in doing so, he raised one of his arms while keeping its elbow on the armrest, and put his thumb and forefinger across his chin. For the briefest moment, thoughts of his ex-wife and his children invaded his thoughts. Memories of his past were constantly intruding upon his consciousness; it was a symptom of his damaged condition. He quickly suppressed the memory, but then something deeper from his subconscious began to surface and replace those thoughts: recollections of that last operation in the Middle East, the one that almost did him in, and was ultimately the impetus for his being removed from his position as a CIA field operative.

Seavers was waiting and could see that he had lost Angstrom; that his thoughts were elsewhere, just like the others had reported seeing whenever they interacted with the man. The Associate Director-level track was definitely out of the question; that fact was made abundantly clear to Seavers at that point.

But over the years Seavers had developed a keen ability to read people, and he did not think that Angstrom really wanted to be groomed for such a position anyway. He also did not think Angstrom's request was some kind of a joke; not after this meeting. The request was a signal, a sort of reaching out for something, and he decided that he was going to seize the opportunity. He had an idea that he had been thinking about for some time. It was a bold decision, but Seavers knew that he could not allow the situation with Angstrom to continue much longer the way it was. Two years of *floating* through his Office and pushing papers on small, meaningless tasks was enough. If he was going to have Angstrom for the long haul, he might as well put him to good use, and in the process, try to help the man heal.

For the longest time, Seavers could not figure out why the powers-that-be dropped Angstrom into the TTO in the first place. There must have been a reason, because there were a whole host of other places Angstrom could have landed instead of his office. It was not until a few weeks ago, while Seavers was thinking about it and preparing for the present conversation, when he finally came to believe that he had figured it out. Of all of the information about Angstrom's background that was redacted in his personnel file, one piece of information that survived the redaction process was his education: a PhD in Computer Science from Carnegie Mellon University, obtained some twenty-five years ago. Why someone with that kind of brainpower ended up working as a field operative was beyond Seavers, but the fact of the matter was that he was now in Seavers' charge, and most likely because it was determined that his PhD could be put to good use there in the TTO.

Seavers' analysis was essentially correct, with one additional aspect. Angstrom was also placed in Seavers' charge because of the high regard held for Seavers himself. He had been around for a long time and was recognized as someone that was well-seasoned at dealing with people; he was good at it. He could hold his own and be firm when necessary (he could not have risen to the level that he did if that were not the case), but he was also a man with compassion.

In any event, Seavers was intent on making use of whatever was left of Angstrom's brain cells.

"Anyway, I have a new assignment for you," he finally said after some time had passed.

Angstrom became alert at the words "new assignment" and straightened up in his chair. He had not heard those words in a long time, and they woke him from his hypnotic trance. His eyes appeared more lucid, and life seemed to return to him.

"What is it?" he said.

A corner of Seavers' mouth turned upwards in amusement as he noted the sudden eagerness in Angstrom, and he was pleased that he seemed to have struck a nerve. His inclination was correct: Angstrom's request to be put on the management track was really not a request for that at all. It was an indication that he was ready—ready, perhaps, to return to the world, or at least, to escape the mind-numbing paperwork that he had been doing. Now the question was: Would what he had to offer Angstrom be enough to intrigue him; was it something that he could sink his teeth into?

"Well, you may or may not know this, but about three years ago our Office kicked off a special effort: the crowdsourced design of a next generation Marine Amphibious Vehicle, or "MAV" as we're so fond of referring to it." He watched Angstrom for a moment to gauge his reaction to what had just been said. The proposition of a government-directed, large scale, crowdsourced design for a military vehicle was a significant, groundbreaking event for the U.S. Military. "Do you know what crowdsourcing is?"

"Yes. It's something like...sending a problem, or task, out to the general public to work on it and come up with a solution. Everyone contributes to the cause."

"Exactly. Now, I don't know how much you know about this, so stop me if I'm explaining the obvious, but a MAV is a specialized transport vehicle for the marines, with tank-like wheels and armor, but with a hull formed to enable it to operate in water as well. It's used to transport Marines and their equipment from water onto land." Seavers stopped to see if Angstrom wanted to interrupt him or say anything.

"Yes, I'm familiar with the concept," Angstrom said. He was being truthful, but not completely forthcoming; he had actually been on a MAV before, and on more than one occasion.

Seavers nodded and then continued, "Well, you see, a few years ago the Marines determined that the current generation vehicle was getting long in the tooth, and that they had to develop a next-generation vehicle with upgraded technology and capability. RFPs were sent out to all of the major defense contractors, and good responses were received. The only problem was that the bids were way out of whack budget-wise. Appropriations would never have agreed to fund the project. So the Marines were stuck. That's where DARPA, and our Office in particular, entered the picture. Our Director had been thinking about trying to crowdsource a large-scale project for some time. We had already tested the crowdsourcing process on a smaller project, but not on anything even remotely as complex as the design and development of an amphibious vehicle. As a result, we created a new program for it. Needless to say, this is a major program with large teams of people involved in a lot of different capacities. I'm sure you've even heard about it in some of the work that you've been doing around here for the last couple of years."

After a pause to let everything register with Angstrom, Seavers continued, "Now, there are actually three crowdsourcing contests, as we refer to them, under the program; namely..." Seavers successively held up his thumb and then two fingers as he counted off, "one: Drive Train, two: Chassis and Armor, and three: Total Platform."

He paused again to let Angstrom digest the information, and then he continued, "When we officially announced the effort two years ago, we made it global in nature; qualified entities from all over the world could participate. A whole eco-structure was created within the TTO to manage the solicitations, including collaboration servers where respondents could modify their submissions real-time in response to further disclosures of information by us and by other contributors. It's all very complicated; there were a lot of preliminary issues that had to be dealt with, many of them matters of first impression. In a way, we've been making up the rules as we go along. New policies on confidentiality, government rights, open source hardware and design issues...it's massive," he said as he held out his hands to emphasize the point. "Sometimes I think that designing the infrastructure within DARPA to manage it all is as complex as the design of a MAV itself."

Seavers got up and poked his head out of the door to let his secretary know that he wanted a cup of coffee. He asked Angstrom if he wanted one as well, but Angstrom declined.

"You're asking the general public, ordinary citizens, to contribute to the electro-mechanical design and development of a highly complex military vehicle?" Angstrom said when Seavers sat back down. "I'm sorry, but that seems really far-fetched."

"Yes, well, to be honest with you, I'm a bit dubious of the program myself," Seavers conceded.

"What does the average Joe know about this kind of sophisticated technology? The guy down the street that runs a lawnmower repair business is supposed to somehow contribute something of value?"

Seavers was pleased; Angstrom was engaging.

"Well, let's not go too far. You're right in that a great percentage of the general public won't have anything to contribute to such a highly technical area as this, even though the whole process *is* technically open to anyone who registers in our system and wants to make a contribution; at least for the first phase. Nonetheless," Seavers said as he raised his right hand outward, "it's something that DARPA wants to pursue, and so we are. In the end, the submissions we received from the big defense contractors will likely carry the day, and the other submissions won't amount to much. So be it." Seavers shrugged his shoulders after he said that.

The secretary came in to hand Seavers his coffee. He took a sip and used the interruption as an opportunity to make a transition in their dialogue. "The deadline for the submissions has passed, and for the last several months, teams of people have been categorizing and organizing the submitted material. So now we've got, for the Chassis and Armor category alone, reams of submissions. Like I said, other than the big defense contractors, and maybe a few university submissions, it'll probably all not amount to much. In any event, we have to analyze it, all of it, and announce baseline winners for each of the three categories."

"There are going to be winners? How can there be winners if it's a crowdsourcing effort?"

Once again Seavers appreciated the level of interest that Angstrom was expressing. "Good point. You're right to some extent, but we're expecting that there will be a lot of major submissions, and ultimately one of them will stand out the most and be picked to serve as a baseline. That baseline will then be further tweaked and modified based upon the other crowdsourced submissions, as part of a next phase of the crowdsourcing effort: Phase-Two. You see, we're going to do this in successive phases, and after the initial selection phase, the winning design will be made available to a select set of registered participants so that it can be modified through additional collaboration. We've got financial incentives attached to certain levels of activity and submission content that meet specific criteria, and we've also funded certain design and development activities directly. You'll see, there are special servers set up to facilitate all of this collaboration. It's quite remarkable."

Seavers reached into his desk drawer and pulled out a thick, vanilla folder and tossed it onto the desk, right in front of Angstrom. "That's where you come in. You're going to lead a small part of that effort for the Chassis and Armor category."

Upon hearing that last part, Angstrom was not initially thrilled at the prospect. It sounded to him like just another paper pushing job, only on a larger scale. He maintained his composure, however, and all he said in reply was, "Lead?"

"Yes, lead," Seavers responded. Then he nodded at the vanilla folder on the desk and said, "You'll be a Program Manager, working with a team of three other people. Their profiles are in that folder. First, there's Andersen Keplar. He's bright. Just transferred into my Office."

"Voluntarily?" Angstrom said in a rather sarcastic tone that he immediately regretted. He told himself that if he was going to finally start living again, including at work, he was going to have to suppress his cynicism. He reached over and picked up the file and began to survey the documents inside of it.

"Yes, voluntarily," Seavers said with a grin on his face, amused at what he took as a form of dry humor.

"Next, there's Fred Book. You may have noticed him, or even interacted with him, in some of the things you've been doing around here." He paused to check Angstrom's response, and after Angstrom shook his head to indicate no, Seavers continued, "He's been around for a long time, and, quite frankly, close to retirement. To put it bluntly, Fred will be Fred, and you'll see what I mean by that very quickly." Seavers took a sip of his coffee. "And then there's Doctor Susan Rand," he said with a subtle difference in tone that Angstrom did not fail to detect. "She's not part of DARPA."

Angstrom already had his head down and was perusing the contents of the file when Seavers said that. He looked up, but Seavers did not immediately say anything else.

"You mean a civilian?" Angstrom asked.

"Yes. She's a PhD from a military think-tank called Anterton Associates, based out of Boston," Seavers responded as Angstrom returned his attention to the file and found a picture of her. "She specializes in manufacturing technology, and she'll be consulting on manufacturing feasibility for the submissions. She does that for DARPA quite often; she has all of the necessary security clearances, and we have a Master Consulting Agreement in place with her. She has a larger role in the overall program, but she'll also be part of your team."

"In what capacity?"

"I'm sorry, come again?"

"In what capacity will she be a part of my team?"

Seavers was pleased with the question. The curiosity was the most life that he had yet seen in him. He took another sip of his coffee. "Chassis and Armor manufacturing feasibility," he answered. "After you've vetted all of the submissions and found the ones that you believe are the strongest candidates, you'll work with Susan to filter through them and determine which ones, if any, have the highest likelihood of being developed and manufactured within budget. In essence, she's going to help pick the baseline winner. She's very smart, and a very interesting person, with quite a magnetic personality."

Angstrom closed the file and tossed it back onto Seavers' desk. This time it was his turn to have a slight grin on his face. The whole concept sounded quite far-fetched to him: DARPA picking the winner of a crowdsourcing contest for the development of a major, new military vehicle for the marines.

His body language did not escape Seavers, who appreciated that Angstrom had his doubts about the whole effort, because he did as well. But, his Office was assigned to manage the effort, and so ultimately it was Seavers' responsibility to treat the program with all seriousness.

"There will be a few other groups of your size that will also be looking at the Chassis and Armor submissions; there are so many submissions that we had to split them up and assign multiple teams to engage in the initial analysis. And of course, other sections of the TTO, handling the other crowdsourcing categories, will be collaborating with you; the Chassis and Armor portion must be seamlessly integrated with the rest of the other MAV categories and design aspects. It's extremely important to define up front all of the interfaces and ensure that there's seamless interoperability of all of the sections. You'll begin to appreciate very quickly the complexity of this vehicle, and the enormity of the program—the hardware, software, materials, and electronics…it's immense," Seavers said.

Sensitive to the fact that he was dropping a lot of information on Angstrom all at once, he paused momentarily to give Angstrom the opportunity to absorb it all. Then he said, "As a Program Manager, you'll need to start attending regular team meetings with the other Managers. They occur once every two weeks; I'll forward the details to you. Susan will also be present at them. You'll find all of the details of the program, more than you'll ever need, in the internal Document Repository under the ZXX-10-09 MAV project designation on the secure servers. Eventually, we'll pick the strongest candidate, and then make a complete and detailed presentation to the Director of DARPA, as well as to a Lieutenant Colonel in the Marines. Ultimately, it'll be submitted to Appropriations for consideration, but that's a long ways away."

Seavers paused to gauge Angstrom's reaction, and then he asked, "Well, what do you think?"

"It sounds like I'm parachuting into something huge…overwhelming," Angstrom said. He thought for a moment, and then he said, "What kind of time frame are we looking at for this."

"Good question. You do have some time, including some padding in the schedule to help you and your team catch up to where things are. In addition, you'll find Susan to be a big help. She's incredibly smart and has been involved with this program ever since its inception. Her firm, Anterton, helped architect many aspects of the program itself. You'll also find all of the budget, travel, and other details in that file," Seavers said as he pointed to the file in Angstrom's hand. "And of course, you can always come to me with any questions or concerns you may have. Consider my door always open. That's it. This is something that you should really be able to wrap your arms around, John, and I'm looking forward to seeing the results."

"I understand." There was a pause before anything else was said. "Thanks, Bill," Angstrom said, looking intently at Seavers for a moment. Then he stood up, extended his hand to shake Seavers', and said, "Looks like I'll have to dust off the old cobwebs."

As Angstrom walked down the hall back to his office, he was not sure how he felt about the assignment, but he felt good about the change it represented. He knew that what Seavers had said was true: he was just going through the motions. For two years he had been moving through the halls of the TTO like a zombie. It was difficult for him not to. Compared to the life that he used to lead, he was a fish out of water working within DARPA. Not to mention the fact that he was dealing with certain internal demons—disturbances in his mind.

He considered the folder that Seavers had given to him. It was thick. It might as well have been put into a large binder there was so much documentation in it. As he walked back to his office, he convinced himself that he would embrace the new opportunity.

In the meantime, today was significant for another reason. He was going to have dinner with his ex-wife.

* * *

It was not apparent to Angstrom why his ex-wife, Breann Evers (she had reverted to using her maiden name), wanted him to come over that night. No reason was given during their brief telephone conversation, and he did not ask for one. She caught him off guard when she called, and full well knowing that he could have used a diversion in his life, there was no hesitation in his accepting the offer. Now that it was time to actually see her, however, he was filled with trepidation. There were a lot of bad memories between them, and a lot of hurt. They had barely spoken to each other since their divorce two years ago, and this was also going to be the first time that he would be in his old house since then.

After he parked his car on the tree-lined street of his old neighborhood, he stayed in his car, leaned over to peer through the window, and looked out at the house. A lot of water under the bridge, he thought. He was nervous, because he knew that she was still a weak spot for him. There was something about her that always made it so.

His failed marriage with her came to the forefront of his thoughts, and like so many times before, he began to analyze how it may have gone wrong. Oftentimes his self-introspection resulted in him blaming himself, primarily because he thought that he could have been different. As tough and in control of matters as he was during his missions, he was a completely different person on the home front—gentle, somewhat meek, and even passive. Being at home with her and the family brought that out in him. If he could have been more of his truer self, as he saw it, more confident, more engaged, and even more worldly and exciting, maybe she would have remained interested in him. Instead, he would become another person. What was it? Exhaustion? A need to decompress? Some type of overcompensation to please her? Every time he returned home from a trip, he would turn into someone he did not know; sometimes he could feel it happening and still could not seem to stop it.

As a result, the initial excitement of their physical attraction, and their whirlwind trips around the world when he was courting her, waned and gave way to the normalcy of marriage.

It was a downward spiral. When things started to go bad, each of them began to believe the worst in the other during any particular situation. They would jump to conclusions, and there was a lot of mistrust.

Seldom would she ever open up and share her feelings. He *never* did. His exhaustion from overseas assignments was compounded by feelings of exhaustion he experienced on the home front, and on those rare occasions when she *did* try to talk to him (perhaps during fits of loneliness or desperation), he was too preoccupied with decompressing from his trip to listen to her, until eventually, she stopped trying. The focus of their conversations, whenever they were together, turned to the negative aspects of their marriage, and that focus crowded out and choked off anything positive that there may have been between them.

Sometimes, when he looked back on it all as objectively as he could, he would wonder whether he was searching for answers when there were none, and that the problem was there from the beginning, at the very onset of their marriage, and when they first met. As his theory went, they each had set a goal for themselves to find someone to marry, a soul mate, by a certain point in their life, and that goal blinded their ability to see things clearly, to realize that they were not meant for each other, or that they were not even compatible. They ignored their instincts; otherwise, they would have quickly noticed certain unattractive, irreconcilable traits of the other. When the romance of the courting was over, so were their attractions for each other. He could never say for sure whether his theory was correct, but he always wondered about it.

One thing that he was certain of was that their situation was exacerbated by his constant travel overseas; he was often away from the family for long stretches of time. Whenever he was home, he could not talk about where he had been or what he had been doing, which increased the level of distrust between them. And where he needed caring and understanding upon his return home, someone to rejuvenate him, and allow him to separate from the stresses associated with his mission, she was anything but. One time, at the height of one of their fights, she yelled at him that she was not a motel where he could just stop in and get some rest before he was off to his next destination. Absence did not make the heart grow fonder for them.

It reached the point where, whenever he was home, he was a disruption to the family. No one, most especially Breann, was used to his being around, and every time he came home, there was a necessary period of adjustment.

Nothing was too small for them to argue about, and they wore each other out. There were too many years of dysfunction to overcome; a pattern of gloom when they were together that neither of them could break. Even before their two daughters were in high school, they came to terms with the fact that they could no longer go on like they were, and that divorce was inevitable. They agreed to stay together long enough to raise the girls, but they had begun serious discussion of divorce when their youngest daughter was a senior in high school. Still, it was a shock to him when no sooner had she graduated that he was served with divorce papers. It could not have happened at a worse time; it was while he was still recovering after having been pulled from the field because of the disastrous results of his last mission.

A motorcycle came barreling down the street, kicking up slush as it sped by, which caused Angstrom to stir from his self-reflection. He was jarred back to the present, and he admitted to himself that there was one thing that he periodically *did* find himself missing since their divorce: sex with Breann.

He could not help it; he still occasionally craved her flesh. And she had a voracious appetite for it. Breann was the only woman that ever controlled him in bed. He used to *like* being controlled by her. It was an oxymoron of sorts, being in such control of matters out in the field during his missions, while essentially being her sexual slave in bed when at home. The sexual acts that he did for her might as well have been straight out of the red light district of Amsterdam.

Maybe it was because he had to be in such control of situations on the job that he liked to be controlled by her in bed. The Psychologist assigned to repair his shattered psyche after that last, fateful mission never came to that conclusion, but then again, Angstrom never disclosed his sexual exploits with his wife to the doctor, and he was never that way with the women that he encountered during his assignments.

When she opened the door for him that evening, he could not believe how gorgeous she still looked. She was as physically fit as ever, and at forty-three years old, she looked stunning, even if her long, full-bodied red hair required a little dye to maintain its vibrancy now.

"Hello John, come," she said while holding the door open for him and with a sensuous look on her face.

What a greeting, he thought; a loaded invitation. He immediately doubted whether he should be there and wondered what he was getting himself into.

He was torn: part of him wanted to lean over and give her a kiss on the cheek as he entered, but the other part dared not, because after all, they were divorced now. He could smell her perfume as he gave her a small hug and his face came close to her neck. It was the same brand that she used to wear for him, and that used to drive him crazy.

"Well, you haven't changed the place much, have you?" he said as he looked around while removing his coat and handing it to her.

"No, not yet at least. I've thought about having the place

painted on more than one occasion, but I just haven't gotten around to it." She hung his coat in the hall closet.

He laughed to himself and wondered what could be occupying her time so much that she could not even arrange to have the place painted, because she was living on his alimony payments—that and the special stipend the Government provided as recompense for the psychological trauma that he suffered—so she did not need to work. "I'm sure it would look fabulous, however you ended up doing it," was all he could think to say.

She stopped to look at him, right there in the hallway and with a smile on her face, like she knew at that moment that she could still control him—still have him—if she wanted to. The smile on his face disappeared almost simultaneously with the appearance of hers, because he realized that somehow he had let it be known as to his enduring physical desire for her. He wasn't sure what he had done to tip her off; it was the same as always in that he wasn't himself with her, and his powers of self control and observation were weakened.

She led him into the living room, and as he followed her, he looked at her body from behind and could not help from lusting after her. Her hips and legs were just as he remembered them when she used to spread her legs and command his services.

As they walked toward the living room she said, "There's a bottle of wine over on the kitchen counter. Would you like some?"

He stopped in his tracks, watched her as she walked to the couch and sat down, and then he said, "Breann, what's this about? Why did you invite me here?"

"John, relax. We have two daughters in college, and we'll always have them as a bond between us. I just think that after all this time we ought to be a little closer than we have been."

"Closer?"

"Just get the wine," she said with a coy expression on her face and a wave of her hand.

He could still remember where everything was in the house, and so in addition to the bottle of wine, he got two wine glasses and went to the couch. He sat down next to her, and as he began to uncork the bottle, Breann said, "So, I've got dinner warming in the oven, and we can eat in a little bit, after we've had a chance to catch up with each other."

After he poured two glasses and handed one to her, she raised hers and said, "Well, here's to our two daughters and their continued success." That was the least controversial toast that she could think of making. What followed were two glasses worth of Breann's summary as to how their girls were doing at college. John listened intently; he was very interested in their progress, as he loved them greatly.

Finally, Breann said, "Let's go eat. I made your favorite roast."

He grew more suspicious when he learned that she had cooked his favorite meal for him. Something was not right; he could feel it. She had some motive in mind and was buttering him up for it, he just knew it. All he could think of was that he hoped she was not going to ask him for more money.

When he sat down at his old place at the table, there was another bottle of red wine sitting there ready to be opened, which he proceeded to do as Breann brought the food to the table. As he watched her move around in the kitchen, he noticed that she was wearing the navy-blue pair of four-inch heels, and the silky blue dress that used to drive him crazy. She must have saved the outfit and shoes for him, because they looked as though they had hardly been worn. He followed the trace of her legs from her ankles all the way up to the back of her thighs, and then to her ass. She knew he was watching her, and as she got the food out of the oven and moved it to the table, she did so in silence and with deliberate, provocative movement, like she was giving him a lap dance.

She tried her best when she sat down to engage in casual conversation with him and act as though everything was normal, like they had still been living together all that time and nothing was bad between them. They took their time and enjoyed eating their meals at a leisurely pace—and then it happened. She was talking about their old friends, and all of a sudden she stopped eating, put her fork down, and looked sideways with an upset look on her face. He had seen that expression before, too many times. Then, just as he expected, she opened up to him and revealed what was bothering her: for over the past year she had been seeing a man fifteen years her junior, and he had just dumped her and ran off with someone his own age. She was humiliated and felt alone. She couldn't face any of her girlfriends, because they all knew about it and were talking about her behind her back.

So that was it. Now he knew why he was there: to console her—console her for the loss of the man that was supposed to have replaced him. She talked about the situation for a long time, and he found himself calming her and comforting her. Why he did so he wasn't sure. Maybe it was because even though he didn't have a lot of compassion for her, he was grateful that she didn't want an increase in her alimony.

By the time they finished dinner, they had also finished the second bottle of wine, and they went back to the couch again. It did not take too long before she took control of him.

"I need you," she said, and moved closer to him and started kissing his neck. He did not resist, but it was all he could do to prevent himself from reciprocating. His inner strength diminished, and he put his hand on the back of her head as she continued to kiss him. Then, after several minutes, she stood up, took his hand, and led him to the bedroom. The alcohol had heightened their desires for each other. He had been abstaining from alcohol for so long that the wine he drank that evening had a pronounced effect on him. When they reached the bedroom, she sat down on the bed, pulled him in front of her, and began to unbuckle his belt. As she began to fondle him, she looked up into his eyes and said in a commanding voice, "I want you to do that thing to me." Both of her eyebrows rose as she said that, causing her to have a sinister look while she gazed up at him.

It was a little after four in the morning when he woke up in her bed. He could not believe what he had done. He laid there in the dark for a while and just stared at the ceiling. Then he sat up on the side of the bed and looked around the room in the darkness. Her silk dress and pumps lay on the floor at his feet. He stayed there for several minutes, thinking about what had just happened. It was not the same for him anymore; not physically fulfilling. He was not mad at her; there was no anger or resentment at her for having lured him into doing what he had just done. Much to the contrary, he was glad, because it liberated him. Her power over him had been severed right at that moment. He looked back over his shoulder to where she was sleeping and watched her for a moment. He had just gone through the motions with her. It was like an out-of-body experience where he had no real pleasure in what had just transpired. It was an incredibly cathartic ordeal for him; it was significant in its own right, and he knew it.

Eventually he got up, dressed, and left the house without even waking her. That was it, he thought. Their divorce may have been official two years ago, but it had become final that night. He was sure of it, and there would be no going back.

Significant aspects of his life were changing. First there was the new assignment with Bill Seavers at the TTO, and now there was the emancipation he was experiencing after having sex with his ex-wife. He was aware of something happening within him; important changes were occurring, and a confidence was slowly returning. Bit by bit, he was coming alive—healing.

When he got into his car to drive back to his apartment, he looked down and saw Seavers' thick vanilla folder sitting on the passenger seat. It was careless of him to leave it there, and he was disappointed in himself for the slip-up. He could not allow himself be so careless and sloppy. He reminded himself of that. The car sped away, back toward his apartment, but he was wide awake and in no mood to go home. When he spotted an all-night diner on the way, he pulled into it. He took the folder in with him, and he studied it, all of it, into the dawn. Sometimes the weekend comes just at the right time.

3

Introductions Overwhelming

A secure data room was assigned to Angstrom and his team to review the Chassis and Armor submissions. A badge had to be swiped across a magnetic reader in order to unlock the door, and only Angstrom's team members were coded into the system for access. When Angstrom entered it on Monday morning some time after nine o'clock, he was not prepared for what he found. Documents, packages, even mini-prototype models, were everywhere, each representing a Chassis and Armor submission. He stood at the opened door and continued to survey the room, startled at the amount of material strewn everywhere, and as he scanned the room, his attention was drawn to a computer with an empty chair in front of it, and then to another computer. Then he noticed a young man sitting at a third computer terminal. As soon as the man realized that someone had entered the room, he turned to look, and with a big smile, stood up and asked, "John Angstrom?"

"Yes, that's right; and you must be Andersen Keplar."

"The one and only. Nice to meet you."

There was an energy about Keplar that was immediately noticeable. It was youthful exuberance, coupled with his excitement for the first day of his new job. And as for Angstrom himself, Keplar was also surprised by his appearance—even somewhat intimidated. Angstrom would be managing him, which itself attached some significance to Angstrom's stature in Keplar's eyes. But Angstrom also just struck an impressive, physical presence upon first impression. They took the opportunity that morning to get to know one another and warm up to each other. Each found that it was easy to converse with the other, even despite their age difference, which was not trivial; they were a good match, and it became readily apparent to each of them that it was going to be easy to get along with the other person.

With blond hair, youngish face, and being of average height, Keplar was just as he was described in the file. He had been with the Department of Defense for one and a half years, ever since he graduated from Yale University's College of Engineering. He had already grown tired of what he was doing within the DoD, and that was why he transferred into the TTO. Angstrom had also noted from the file that both of Keplar's parents served in another branch of the DoD. It was an interesting development, he had thought upon learning it—Keplar was connected.

After they had spent some time getting to know each other, Keplar, full of enthusiasm, looked around the room and said, "Well, given that this is my first day on the job, I thought I'd get an early start of it, and so I came in early this morning to start reviewing the material."

"Hmm. And how's it going so far?" Angstrom said, noting Keplar's reference to the material lying about in the room. Before Keplar had a chance to answer, Angstrom picked up a little remote-controlled car, one that was readily available at stores but that had been crudely modified with the intention of serving as a contest submission, and while he studied it he said, "What's this, some kind of a joke?"

Keplar laughed and said, "No, it's not...and then again, yeah, it is. I mean, you should see some of the *shit* I've looked at already. It's gibberish...real gibberish." He paused for a moment when Angstrom turned to look at him, and then he said, "Sorry about the language."

"Hey, don't worry about it...not around me."

Angstrom recalled how Seavers had told him that for this first phase, the whole crowdsourced effort had been opened to a large number of registered applicants, and that at a later date, during subsequent phases, the process would become more restricted and closed to those entities that did not meet certain minimum requirements and criteria through the official registration process. So the floodgates were open. He looked over Keplar's shoulder at the computer beyond him to try and see what he had been doing. "What do you have going on there?"

"Well, we have all kinds of submissions to analyze," Keplar said as he waived his hand in reference to the submissions throughout the room. "Paper ones, mechanical samples, electronic ones...and the electronic submissions are available for review on the computers. The files have already been through an intense scanning process to check for viruses, Trojan Horses, illicit code...you name it."

"It looks like everything in this room has already been opened," Angstrom said as he picked up a large, yellow envelope and looked inside of it.

"Yeah, it has. It's been inspected for bombs, x-rayed, checked for Anthrax, the whole nine yards. At least we know we're safe. But nothing was thrown out or removed based upon merit. It's *all* here...all for our review and pleasure."

"Hmm. Why the latex gloves?" Angstrom said, pointing to the gloves that Keplar was wearing.

"Shit, they gave 'em to us, so I thought I might as well be safe than sorry."

"I see."

Angstrom surveyed the room again as he swiveled in his chair. The magnitude of their task was becoming more apparent to him by the minute. He recalled what Seavers had said about how they might go through all of the submissions and end up with nothing—nothing of significance except for the contributions from the large defense houses, and maybe a few universities. While he was sitting down, Keplar covertly took off his latex gloves, having become self-conscious about them after Angstrom pointed them out. He tried not to let Angstrom see him take them off.

Angstrom took a deep breath as he contemplated the enormity of the task, and he reminded himself that he needed to stay motivated for it. He had to. He needed to finally get out of the funk that he had been in and seize the opportunity Seavers had given him, however great that opportunity might or might not be. He had to come to grips with the fact that his life in the field…that kind of life…was over. He thought that he had already convinced himself of it, but entering the small data room and seeing all of the submissions, piles and piles of them, was like cold water splashed in his face; it was jarring.

"Did you have any process in mind when you started reviewing this stuff?" he finally asked Keplar.

"No, not really. I'm kind of just going where the mood strikes me." Keplar turned to look at a diagram displayed on the computer monitor. "Here, look at this," he said, pointing to the monitor. "Someone cut and pasted a little car from a Mario Kart video game and then used a drawing program to modify it. That's the entire submission. That's it. Ridiculous!"

"If the press ever saw it, they'd have a field day, that's for sure," Angstrom said. "I'm going to start over here in this pile and take a look."

"You can, but I put all of those together in a pile because they're the big defense contractor submissions. My guess is that they're going to take the most time to review, judging by the size of them, so I put them there for later. I guess I'm sort of filtering out the easy stuff first."

Just then the door to the data room opened, and Fred Book walked in. He was thin and frail, and he looked old. His attire harkened back to another era, like it was handed down to him from the engineers of a NASA flight control room back in the fifties. The skin on his face was darkened and leathery from having smoked for so many years. Both Keplar and Angstrom stopped what they were doing and watched Fred as he entered. At first he said nothing to either of them except for a short, "Gentlemen," with a nod of the head as he continued to survey some of the submissions lying around. He was sifting through them like he was looking through his own mail, and after a while he took one of them, walked over to Angstrom, and held it out toward him.

"What? What do you want me to do with this, Fred?" Angstrom asked as he accepted the file from him.

"I'm checking it out?"

"Huh? I'm sorry, I don't understand. By the way, we've never really met before. I'm John Angstrom," he said as he held out his hand. He had seen Book around the office quite often and noticed that he kept to himself and never conversed much with anybody else. Occasionally, he could be seen talking to Seavers in the hallway, but that was about it.

"Uh huh. Nice to meet you," he responded as he shook hands. "Look, I'm not going to sit in here and review this stuff," Book said as he looked around the room. "Official procedure is to have the Program Manager register all materials that are checked out using the online system. I'm taking this material to my desk," he said as he held up a file that he took from one of the piles, "so you have to register it." Book saw the blank expression on Angstrom's face, so he continued, "Think of it like the library: things get checked out, and then they get checked back in." He explained this in an agitated tone, miffed that he had to explain such a widely known procedure to a Program Manager. "Come over to the computer," Book said as he sat down in front of one of them, and I'll walk you through it."

After this was done and Book had left the room, Angstrom and Keplar looked at each other, smiled, and shrugged their shoulders. Then Angstrom turned his attention back to one of the piles of submissions and selected a submission to begin reviewing it.

For the next several hours, they went through submission packet after submission packet. Most of them were easy to dismiss without too much consideration: a tricked-out golf cart; a VW mini-van from the sixties with snow-chains on the tires and a floatation device attached around its perimeter; a large, orange school bus with an aluminum boat somehow attached to its underside. A Jeep with a proposal to modify its axel system, allowing the four wheels to expand outward into a position parallel with the plane of the bottom of the vehicle and thereby serve as a flotation system, held Angstrom's interest for a while, but it was just idle curiosity.

Eventually they both needed a break; the mind can only take so much of the sort of technically dense material that they were exposing themselves to. Being their inaugural workday, they decided to have lunch together in the building's cafeteria, and upon entering the seating area of it, with trays of food in hand, they noticed Seavers sitting alone at a table and joined him. It was a welcome respite for all of them, and Seavers was especially pleased to see Angstrom successfully coming out of his shell.

They resumed their efforts in the data room that afternoon. "By the way, did you read the whole Chassis and Armor bid package?" Keplar stopped and finally asked at one point later in the day, partly to hear Angstrom's answer, and partly to just to break up the monotony.

"No, not all of it. But I read some of it over the weekend," Angstrom responded. "You know, it's odd that we're supposed to be reviewing submissions for the Chassis and Armor category, yet I'm finding responses here that are for the entire vehicle design, with nothing specific to Chassis and Armor for the most part."

"I know," Keplar said. I think that's a reflection of the quality of the submissions. The submitters can't even follow directions or understand the basic task. But not all of them are that bad; some of them are on point. I think I'm going to spend the next couple of days reviewing this one proposal," Keplar said as he held up a submission by one of the major defense contractors, "just to get an idea of what a serious Chassis and Armor submission is supposed to look like."

"Probably not a bad idea," Angstrom agreed. "Do me a favor: hand me that packet over there and I'll brush up the same way."

Keplar handed him another one of the submissions, and as he did that he said, "It's kind of odd that they're assigning people so fresh to the technology, like us, to do something as important as reviewing these submissions."

Angstrom did not respond to his statement, but he thought about what Keplar had said. It *was* odd. They did not know the first thing about the intricacies of MAV chassis and armor technology.

Upon further reflection of what he was about to do, Angstrom had a change of heart and decided to go back to the original Chassis and Armor bid package, the one that was made public two years ago, and study that first, and then after that he would resume his review of the large defense company packages. It was going to be an immersion, a deep dive, so to speak, into chassis and armor MAV technology.

He went onto the computer system and opened the electronic version of the bid package stored on the secure server system. The first major section was the "Objectives and Goals." There was page after page of detail: "*5000 - 12,000 total parts...minimization of thermal, electromagnetic and vibrational emissions...survivability requirements...optional use of virtual vehicle design environment ...novel foundry build compatibility required...novel (e.g. v-hull) configurations encouraged....*"

After a couple of hours, he struggled to keep his eyes open and stay awake because the material was so dense and exhaustive, and he decided to take a break, stretch his legs, and get himself some coffee. As he walked to the machine, he thought again about what he was getting himself into; the whole MAV program was a massive effort; Seavers was right about that. Angstrom was just reviewing the Chassis and Armor submissions—one aspect of the overall program—and the complexity of that alone was enormous.

They studied submissions in the data room all day, and that first day felt like a marathon to them. They were not used to pacing themselves yet. At around six o'clock they both had enough. "That's it for me," Keplar suddenly said. "No use having all the fun in just one day." He slammed the binder shut that he had been reading and stood up to stretch.

"Sounds good to me," Angstrom responded. He was mentally exhausted as well.

"I'm going to go shoot," Keplar said as he gathered his things and prepared to leave.

Angstrom looked at him with a puzzled expression on his face. "*Shoot*? What do you mean 'Shoot'?"

"I signed up for lessons at our shooting range; an information packet offering the lessons was included with my job offer, and I figured, what the hell, why not? Maybe it will help me blow off some steam or something. I'm really pumped about it."

"Hmm. Interesting."

"Yeah. They're starting it off in high gear, too."

"What do you mean? How are they doing that?"

"The syllabus said that we'll be meeting at what's called an extended range to fire an AN-94."

"Are you kidding?" Angstrom said with surprise. "Why the hell are they letting you guys play with a Russian assault rifle?" He was really just thinking aloud more than posing the question to Keplar.

"You know what an AN-94 is, eh?"

"Yes, I do. My question is: What's the purpose of letting a new-hire like yourself play with one?"

"I don't know. Hey, I just signed up. Maybe it's a way for us to stay on top of the technology; you know, since that's what we do: analyze and develop military technology," Keplar responded. After a brief moment he said, "You want to come with me?"

Angstrom was caught off guard by the offer, but he considered it. It was something that he thought he would never get close to again—firearms. He wondered what the doctors would think if they knew that he was going to pick up an AN-94 and fire it.

"Yes. Yes I do," he finally decided.

"Great, because to tell you the truth, I wasn't looking forward to going alone."

They both started to walk out of the room, and then Angstrom stopped and said, "But I get to pick the place where we have dinner afterwards, right?"

"Sure. You got it. You buying?"

"No, I'm not, and don't get any idea that I'm loaded or something like that, because I'm not. What time does it start?"

"In about a half-hour. I'm going straight over. Do you want to ride with me?"

"No, let's go in separate cars."

"Okay, sure." While they were walking in the hallway, Keplar said, "Wait, does your car have entry rights? It's a highly restricted area. I had to make a special application for my car to have entry rights when I signed up for the course."

"Don't worry about it. Listen, try not to go through any yellow lights on the way, but if we get separated, I'll meet you over there. Okay?"

"Sounds good." They proceeded through the electronic revolving door to exit the building, Keplar going first. When they were both outside and began to walk to their cars, Keplar said, "You know where the range is, huh?"

Angstrom nodded his head but did not say anything else.

Interesting, Keplar thought.

Upon arriving at the special assault rifle test range, they each swiped their badges to gain entry into the restricted area. The place was teeming with people that evening. Keplar looked down at his invitation to see where they should go, and he led them through the area like he was following a map to a treasure. Angstrom was very familiar with the area and knew exactly where they were going when Keplar told him the location, but he let Keplar do the leading. Eventually Keplar found the location, and as they walked toward the stall, they could tell that the session had already started, because there was a small group of people gathered and listening to the speaker's instructions.

"This is a Russian AN-94, or more precisely, the Automat Nikonova," the instructor said as he held up the weapon. "Its recoil force is delayed until the fired round has exited the barrel...."

Angstrom drifted from the crowd and began to walk around the range. It had been a long time since he was in the cave, as it was commonly referred to. He reminisced about the many times he had been there, training on different weaponry, and the many people that he had associated with over the years. He and Gordon used to spend a lot of time down there together, honing their skills. It felt like an eternity to him since he had last been there.

After a while he snapped out of it and made his way back to the small group. Each person seemed so young to him; they looked like kids handling the AN-94, and they treated it so gingerly, like it was made of gold.

"What about you, sir? You're the last up; care to fire?" the instructor, an older man, asked Angstrom, who by that time was standing behind everyone else.

He froze for a moment at the invitation. It would be a big step for him. But he had already come as far as he had, and he knew deep down that he had intended all along to fire the weapon. And so, he adjusted his ear protection and stepped forward to receive the weapon. The other people in the group, including Keplar, did not pay any attention; they had already moved backwards, some distance behind the spectator line, and were distracted by their own conversations with each other. The man who was about to fire the weapon did not hold their interest. Angstrom looked out of place next to all of them; a grandfather among children.

He stood and looked at the small target far off in the distance, planted his feet, slowly brought the sight up to his eye, and fired two quick bursts. The power of the gun, and the sound it emitted upon firing (even after being attenuated by his ear protection) felt good to him. A target indicator immediately emitted a high-pitched noise, and a red light flashed twice, indicating that both rounds hit the center of the target. Everyone turned to look at him in surprise; no one else had come close to hitting the target, much less setting off the bulls-eye indicator.

The old man who was the instructor had a slight grin on his face as he received the weapon back from Angstrom, and he said rather quietly, so that the others could not hear, "I thought that was you." Then he turned to the rest of the group and said in his normal voice, "That, ladies and gentlemen, is a perfect example of the firing of an AN-94 Russian assault rifle."

As the instructor pointed out the correct things that Angstrom had done in firing the weapon, Angstrom peeled away from the group and began to walk toward the firing range exit. The lesson was not over, but he had had enough and was ready to leave. Keplar caught up with him and said very excitedly, "Pretty impressive. Where'd you learn to shoot like that?"

"I was a co-leader in my daughters' Girl Scout troop," he said as he walked through the revolving glass door.

"You're leaving?" Keplar asked in surprise through the glass door. "It's not over yet."

"It is for me. You stay if you want. I'm going."

"Okay, okay, wait. I'll come too."

As they walked out of the building together, Angstrom said, "Remember, I get to pick where we're having dinner." He proceeded to give Keplar the name of the place and directions to it in case their cars got separated in traffic. Keplar was bewildered at their sudden departure from the range, and he had to walk fast to keep up with Angstrom as they walked through the parking garage.

After they were seated at their table at the restaurant, Keplar looked around the place and said, "Well, this isn't what I was expecting."

"Why? What were you expecting?"

"I don't know, maybe a dark, wood-paneled steak place."

"You can get a good steak here, if that's what you want. They grill excellent steaks."

Keplar was beginning to notice, after spending only a day with Angstrom, that he was a very intense, serious person. "You're not a steak person, I take it," he said.

"No, not in a long time. Mostly fish for me anymore, for health reasons."

Keplar ordered a Guinness, and Angstrom, telling himself that he had had his share of alcohol at that dinner with his ex-wife, ordered a bottle of mineral water.

"Not drinking either, eh?" Keplar said with a grin. "Come on, you're not *that* old." When Angstrom did not respond, Keplar switched topics and said, "So, now that we're away from the rest of the class, tell me: Where *did* you learn to shoot like that?"

"Sorry, can't talk about it," Angstrom replied. But as soon as he said that he regretted his answer. It was too abrupt and evasive, and he felt like he needed to say more. If he didn't, he knew his answer would not sit well with Keplar, and that he would become even more curious.

The waitress came and put a basket of bread on their table. After she left, Angstrom said, "That may have sounded evasive, and I guess it was. I realize that we're going to be working together a lot, and you're bound to want to know more about me, which is only fair. The thing is: I'm a pretty private person." He paused to think of what else he could say. It was the same issue that he always faced whenever someone inquired about his background. And he came to the same conclusion; even though he would like to say more, he knew he couldn't. Whenever he opened the door on the subject, even slightly, the conversation would go in all kinds of directions, thereby making the situation even more tenuous. Therefore, he needed to bite the bullet and just cut off any further line of inquiry on the subject. "I'm sorry, but that's just the way it is. I don't like to talk about myself. It's nothing personal, and I hope you can understand."

Keplar nodded his head slightly. As Angstrom reached for some bread, Keplar turned to look at some of the other patrons in the restaurant, wondering what he was getting himself into.

When he transferred out of his old job, it was because he was looking for something different to do. Reviewing information in a data room was not quite what he had in mind, and now he was finding out that he was going to be working with someone who sounded like an introvert, or at the least, a little stand-offish.

Angstrom noted the disappointed and concerned expression on Keplar's face and felt bad that he could not say more, but it was not the first time that he saw such a reaction by someone. He was used to it. So, he did what he always did and changed the subject. "You got an engineering degree from Yale; very impressive. I understand that Yale has an outstanding program."

Keplar nodded without saying anything. Angstrom could tell that he had not broken down the new barrier that had just formed between them. He had to work more.

"I'm curious: I saw in your file that you're looking for work that's more technical in nature," Angstrom said as he bit into a wheat roll. "Why did you join the DoD in the first place? You could've gone to a lot of high-tech companies instead, and probably gotten paid a lot more."

Keplar's knee began to bounce up and down nervously under the table, which did not go unnoticed by Angstrom. "Well, to be honest with you, my parents had a lot to do with it. Being in the government themselves, they have high aspirations for me for a government career. I suppose I do, too." Angstrom somewhat bristled at the response, without Keplar noticing. The concept of an ambition for a *government career*, in the abstract, seemed like an incredibly odd notion to him.

Keplar paused for a minute to better formulate what he wanted to say next. "I just didn't like working in computer forensics anymore. One and a half years sitting in front of a computer screen was enough of that." He took a drink of his Guinness, wiped a little froth from his upper lip, and said, "I really don't see myself as a true techie forever. I want to build a solid background in it, and then move on."

"*Move on*? To what...teaching?"

"Shit no. Administration."

"*Administration*? Are you kidding me?" Angstrom said, slightly exasperated. He recalled seeing something to that affect in the file, but it did not register with him until just then. "What kind of Administration?"

"I don't know. I'm not sure yet. Something in the government. DARPA... DoD... maybe even the CIA or NSA. Who knows?"

"I see." He wondered if that was really Keplar talking, or his parents. Who has aspirations at such a young age to work in Administration?

After a time, Angstrom decided to bring the conversation to the more immediate—their project at work—but just as he was about to do so, the waitress brought their dinner orders to the table. She placed a fourteen ounce New York Strip and a baked potato in front of Keplar, and a grilled Salmon with steamed vegetables in front of Angstrom.

"Ah, this looks excellent," Angstrom said as he looked over at Keplar's plate, and then at his own. "Well, why don't we talk about how we're going to progress with the rest of our analysis of the Chassis and Armor submissions? What's your opinion after looking at it for a day? How do you see us tackling it?" Angstrom began slicing into his Salmon.

"Well, I was thinking about that," Keplar said as he began to chew a large piece of steak. "I'm thinking we should quickly go through each of the submissions and do a high level filter of them. As we've already seen after just one day, there's quite a bit of ludicrous stuff—some probably even hoaxes—and we should quickly get rid of that." After swallowing the piece of steak, he took a drink of his Guinness and continued, "Then we'll be left with the more serious submissions. Somehow, we'll have to figure out how to pick the winner from that, which I'm guessing is still going to be a Herculean task."

"Yes, I agree. It's not going to be a walk in the park, but I like your proposal," Angstrom said as he continued eating. "What I'd like to do is have you continue doing the coarse filter of all of the submissions, like you mentioned, and then I'm not even going to bother looking at anything that you've already rejected. While you're doing that, I'm going to focus on some proposals from a couple of the large defense houses. I want to get a better feel for the technology, and what a serious submission looks like. Then I'll be able to review the submissions that survive your filter with a more focused lens. That'll also help make sure that I'm fully up to speed on the technology by the time of the Phase Two pre-meeting that's coming up in a couple of months."

Keplar looked up from his plate and said, "What meeting. What's 'Phase Two?' "

Angstrom proceeded to tell Keplar all about the upcoming meeting, and he decided right then that he wanted to extend an invitation to him, although he supposed that he would need to get approval from Seavers for that. And then he opened up a brief discussion on Susan Rand, and what he knew about her from the file.

"As for Fred," Angstrom said, "...we'll just let Fred do what Fred does." Both of them grinned to each other.

After dinner, they declined the waitress's offer for coffee or desert. Then Angstrom jokingly indicated that it was past his curfew, and that he was ready to leave.

"You go on ahead, John. I saw a couple of interesting looking females sitting over at the bar by themselves, and I'm gonna go try my luck."

Angstrom grabbed the check and rose to take it to the front to pay. "Suit yourself." He looked around the corner to check out the women momentarily, and then he said with a smile, "I'll see you tomorrow."

"You got it...and thanks for dinner. I owe you one," Keplar said with a grin, in recognition of the fact that Angstrom had picked up the check.

Standing at the table, Angstrom looked down at him sternly and said, "No, you don't."

Keplar tilted his head sideways with a puzzled look on his face as he looked up at Angstrom.

"Don't give IOUs so easily," Angstrom said. Then he left.

The smile on Keplar's face disappeared. What a strange

person, he thought—really a strange bird. Then he leaned over to see whether the two women were still at the bar, and a smile returned to his face as put his napkin on the table and stood up to go try his luck.

4

In Rumination of the Deplorable Act

It was frigid outside—the type of Sunday evening that the Professor used to relish. The wind howled as it traveled through the crevices of his old, wood-framed country home. He stood by the fireplace, reached to grab another log from the pile, and threw it into the fire. After a slight hesitation, he grabbed another and tossed it into the flames. He still liked a big fire.

The burning wood began to pop and crackle from the moisture in it, and he sat in his chair near the fire and watched the embers from previously expended logs rise through the flames and cause a glow in the darkness of the study. The wood was chopped from maple trees that were cut down on his land, and the sharp, satisfying aroma emitted by them permeated the air. The warmth from the fire felt good on his old body.

He reached over and clicked on the Halogen floor lamp that arched over his chair and adjusted its intensity; the light was just bright enough for him to read his copy of the *St. Petersburg Times* newspaper, but not so intense that it would disturb the calm ambience of the room. Tchaikovsky's Sixth Symphony, the

Pathétique, was playing in the background; the stereo system, including the turntable playing the album, was so old that the amplifier operated with vacuum tube technology. He had a modern system that he could have used to play such music, but sometimes he still liked to listen to the music in its original, analog format, hearing the pops and ticks from the nicks in the vinyl from years of use, and he liked seeing the vacuum tubes glow orange between the slits of the cabinet vent as the tubes amplified the signals fed through them. It brought back memories of better times—times spent together, with his wife.

As he flipped through the pages of the newspaper, an article about how an electronic, interactive whiteboard was being held at a customs warehouse on Russia's southern border caught his attention. The device's Turkish manufacturer was rushing to get the units into a Moscow showroom in order to start selling the product to schools within the country. The article implied that the whiteboards represented the latest in education technology. An accompanying photo, however, showed an outdated incandescent light and mirror system. The Professor thought that the newspaper must have made a mistake and used the wrong image, because it was not the image of a modern smartboard. Upon closer inspection of the small words below the picture, however, he saw that it referred to the image as an "interactive whiteboard," not a smartboard. He grimaced in disbelief. How could such an archaic projector be referred to as new technology? He figured that some administrator responsible for the procurement of school equipment must have been bribed into buying the outdated product. What made the whole story even more disturbing was the likely reason that the equipment was being held up at customs in the first place: one of the agents had probably not received his cut.

It was nonsense to him, and he went over to the fireplace and tossed the paper into the fire out of disgust. The flames grew and engulfed the paper, quickly turning the white pulp into withering, black debris. Loud, popping noises sounded, as if in commentary to the contents of the fuel being consumed.

It was difficult for him to read the local papers anymore. There was just too much corruption evident when he read between the lines.

He walked over to a bureau in a dark corner of the room, grabbed a snifter, and poured himself a cognac. He was moved by the sounds of the violins playing in the background, and he stood in the darkness for a moment, twirling his glass in order to stimulate the rich aroma of the alcohol. The sound of the flickering flames, mixed with the pure beauty of the symphony, and the bouquet of his elixir, soothed him.

On the way back to his chair, he picked something up from a nearby table and sat back down with it. It was somewhat of a wonder in Russia: a modern tablet computer. They were rare even in St. Petersburg, but he was able to obtain one under the auspices of his duties as a technical consultant to the Russian military establishment. When he awakened it from its sleep and navigated to a certain application, the display revealed something else that was unusual in Russia: unfettered internet access, and to content that originated from outside of his country. Yet another privilege that he enjoyed due to his position.

If the Kremlin knew that he utilized the device to access online newspapers from the United States, simply for the purpose of his own enjoyment, would permission for such access continue? They knew about it by now, given the tight surveillance that was placed on him, and yet the access continued. Those of privilege and rank in Russia operated according to a different set of rules and principles as compared to the ordinary citizens at large.

He was surprised, however, that after all he had been through, and all that was going on, such a luxury still existed for him. It was, in fact, quite suspicious. He certainly was well aware of the fact that his internet usage was being closely monitored. Everything he did now was being watched, tapped, or intercepted; he was a mouse in a cage, constantly being observed. The KGB, or some division of it, whatever the present incarnation of it may now be called (he did not even know anymore), had him in its sights.

As he surfed through the electronic edition of a U.S. newspaper, something curious appeared on the screen: a short news excerpt concerning Moscow. How odd it seemed that on that particular evening, when he was searching for news on something other than his own country and its problems, such a news item would appear. The story indicated that two officers had been sentenced to prison after being convicted for the illegal detention of a resident in the city of Kazan, and for causing that resident's ensuing death. The investigators alleged that the resident was beaten and sexually assaulted by the officers before his eventual death.

It was definitely not the kind of news that the Professor was looking for, and in an attempt to find something else, some article of interest that was far removed from the world in which he was currently a prisoner, he entered the word "education" in the search box to see what he could find. With a tap of his finger on the display, and then a couple more swipes across it, he navigated to an article listed from an older edition of the online paper. Here, perhaps, was something of interest that he could read without causing his stomach to turn. The article's headline was posed as a question: Do Universities in the United States continue to hold their prominent standing in the world of academia? He began to read the article with earnest. It was well written with sound principles and analysis, but there was one glaring omission in everything that he read. The discrepancy was most prominent when he looked at a graphic that accompanied the article: it was a bar graph comparing the total 2010 college-level education expenditure, in billions, for the United States, China, and India. Russian expenditure was not even depicted. He scanned the whole article, and there was no mention of Russia; its educational system did not even merit consideration. The Professor was not sure whether to be angry or sad.

The *Pathétique's* Finale began playing—*Adagio lamentoso.*

He carefully put the tablet down on the small table next to his chair and walked toward the fireplace. The blaze had diminished sufficiently so that he could stand by it without the heat assaulting his body. He rested his arm on the mantel and looked down into the glowing logs. Taken individually, the articles that he had just read would not have meant much. The dispirited state of his soul, however, coupled with the irony represented by the collection of those articles—of all the news that could have been presented that evening, those were the ones that surfaced—made it seem as though the disparate items were collectively telling him something different than what each of them disclosed individually: Russia was greatly damaged and deteriorated, perhaps more so than ever in its history, and seemingly beyond hope.

He tried for a moment to be objective, and wondered whether his own, unique situation, and the corresponding tragedy in his life, skewed his view too much. Of course not, was his answer; it was the conclusion he always came to whenever he harbored such doubt. The spiritually bankrupt state in which he was in was directly proportional, and mirrored in magnitude, the degraded state of his country.

And then, as happened so many times before when he became reflective on the state of his country, and of his own personal situation, he thought of that awful tipping point that occurred in his life—the one that finally brought him to where he is. The single event that transpired, those several years ago, which caused him to become absolutely disconsolate in life. The fire hissed as moisture in the burning logs met flames, as if it were giving sound to the anguish in his soul.

That so called turning point in his life was at a time when he had decided to begin to speak out openly, in University forums, about the critical need for reform in Russia. His wife did not even know that he was doing it at first. When she attended one of his seminars and heard him speaking the way that he was, they talked about it all night in their home; right there in the very room in which he was now sitting. She was proud of him, but warned him that nothing good could possibly come from it. She neither stopped him, however, nor discouraged him. After living together for practically their whole lives, they thought and felt about things in the same way, and she heard the passion in his voice as he spoke about it. They were one soul.

It did not take long before the Defense Ministry took notice of his speeches and asked him to stop. At first they asked him politely, and delicately, but when it continued, they expressed themselves more sternly. Such talk was not allowed, they said, especially by someone in his position. He was biting the hand that fed him; what an embarrassment it was becoming to them.

The Professor did not yield, and when, during a question and answer session after one of his talks, a hostile audience member was reported as having yelled out that the Professor was a *Scientific-technical intelligentsia*, the political leadership decided that things had gone too far. He was inciting the people.

That was when the real pressure began and the KGB became involved. He first noticed it when he began to recognize a certain man that kept appearing at all of his speaking engagements. He would always be there, seated in the same place in the audience: front row, center seat, looking right up at him.

Then the Professor became cognizant of the fact that his house phone was being tapped. The telltale signs were obvious to one such as the Professor, who was so knowledgeable in wiretapping technology.

After that, money was clandestinely siphoned out of his bank account and then mysteriously deposited back again. Sometimes the account balance would be taken to zero and remain there for close to a week. They were warning him of their power; threatening him. He and his wife were followed everywhere; on the streets and in stores, wherever they went. Then his work became impacted; reports that he submitted to the government for his research projects went unacknowledged. Every aspect of his life was being affected. Even his wife began to wonder whether he should just stop, but still she never said anything. He was on a path now, and firmly determined.

His University's leadership tried to discourage him. When that did not work, even his computer at the University was affected; it became sluggish—someone was remotely logged onto it at the same time as he was, monitoring him. Finally, all of the funding for his research dried up. His livelihood was being choked off. Without research funds, he had no compensation, save for the small stipend he received for teaching a few courses each semester.

In reaction to it all, he responded with what he thought of as his own heightened form of retaliation. He completely detached himself from his University's cooperation with the military, withdrawing from, and not helping it respond to, any RFPs for military hardware projects, RFIs, or any other types of solicitations for input.

Why should he lend his mind and intellect to help strengthen those that he thought should be weakened and crushed, and that were against him?

He became isolated from the establishment.

To further his aims, he convinced four of his academic colleagues to join him and redirect their research efforts toward something that was more beneficial to the people of their country: biomedical research and technology.

After this switch, the next step for him, the final, perilous one that was the last straw, was his overt speeches that took direct aim at, and attacked, the Russian Government and its leadership. Whereas up until that point he had spoken in broad terms on the poor state of Russia, and had cast implied aspersions against Russian society generally, now he directly attacked the political establishment.

It was then that it happened—the Deplorable Act, as he thereafter thought of it—when the KGB pressure reached its apex, and propelled him to his spiritual nadir. It was the impetus for his current course of action. That Act alone caused him to realize, with acute clarity, the dire nature of the situation…and the hopelessness of it all.

It was on one evening when he returned home from work. He found his house broken into and ransacked; highly personal belongings, cherished for decades, were destroyed. But the worst of it was what had been done to his wife—that beautiful, old woman, the love of his life and his soul mate in the world. He found her lying on the floor, bound, blindfolded, naked, and sobbing.

After he gently untied her, sat her down on the couch, covered her with a blanket, and gave her something to drink to calm her nerves as best he could, he learned that she had been injected with something.

"Calm down," he tried to say to her. "What do you mean? Slow down. Explain everything to me."

She told him all that she could remember; how the men broke into the house while she was napping, bound and blindfolded her, and dragged her down the stairs before throwing her onto the floor. There she lay as they ripped apart their home. She could discern from what little they said that there were two of them—two men; two *Angels of Death*. All of a sudden, they stopped what they were doing, and there was silence.

She heard their footsteps as they came toward her. One man held her naked body in place—being so frail with old age she could not resist—while the other man forced a needle into her arm; something was injected into her veins. They said nothing to her.

A perfunctory search for evidence was performed later, police reports were filed, but no clues were found, and no one was ever caught.

The psychological trauma that his wife experienced was devastating enough, but that was the least of the harm she suffered. The first signs of the real injury were fever, chills, and night sweats. In the coming weeks she lost her appetite, lost weight, and became bedridden. It was the injection; he knew it.

After extensive testing, she was admitted to the hospital, diagnosed with having a dangerous strain of drug-resistant tuberculosis, and placed in isolation. He was temporarily admitted to the room next to hers and placed in isolation himself as a precautionary measure, until it was verified that he was not infected.

The doctors assured him that the special antibiotics that they would administer would have a good chance of saving her, but her condition only worsened. Her chest hurt when she breathed, and she developed a persistent, deep cough that produced a thick, mucous discharge. Then she started coughing up blood.

Despite all of his protestations, and consultations with physicians from other hospitals, nothing could be done to help her, and the infection spread into the pulmonary artery, resulting in massive bleeding. He plunged into the lower depths of despair as he witnessed the worst that Tuberculosis could wreak on the human body. The pain became unbearable to her; she could not endure it anymore. He cried at her bedside, helpless, gowned and masked to prevent his own infection. Her central nervous system and lymphatic system were failing, and she was in excruciating pain.

It was only due to the mercy of one doctor, who disobeyed orders and agreed to suspend her life support in the middle of the night, that the brutal suffering was finally over. When she breathed her last breadth, and the medical instrumentation indicated that she was dead, he was crushed. He would never be able to erase those haunting, final images of her from his mind, for she had been everything to him, and all that was left of her was an assemblage of flesh, blood, and shredded nerves, and then she was gone.

Right before the end, before his wife took her last breath, a young woman had entered the room and stood by the door, just observing the suffering old woman; she put a hand over the mask that covered her mouth, and tears streamed from her eyes in response to the horror of it all. After witnessing the death, she left without saying anything, keeping her hand over her mouth. No words were uttered.

When the Professor had left the hospital that day, there was a man standing outside of the building, waiting for him. That image was vivid in the Professor's memory as he thought back to it, for the man looked at the Professor with a sinister expression of wickedness. It was the same man he used to notice from a distance when he stood at the podium for his speaking engagements and peered into the audience. His appearance struck the Professor as one of pure evil: long, narrow, bald head, dark, piercing eyes, and a beaked nose. It was the devil himself— *Beak-nose.*

The man did not say anything to the Professor that day; he did not have to. His look alone said it all: *That is what we can do, and what he have done.* The Professor's passport was subsequently revoked, and all routes to travel for him were suspended after that day.

A sudden knock on the back door of the Professor's home disturbed him from his thoughts. Like so many times before, there were tears in his eyes. Anytime he remembered that day, it was heart-wrenching for him. The knock at the door was Dmitri; the Professor was expecting him that evening. He put his cognac down on the mantel, collected himself, grabbed his hat and coat, and went to the door.

They both knew that they could not talk inside of the house because of the eavesdropping. They had to be cautious about even *meeting* at the Professor's house; he was sure that there was surveillance of his home. There was nothing wrong with them meeting for a friendly night of drinks and chess, however, and that was their story.

The real discussion needed to occur outside.

When the Professor opened his back door, a barrage of cold air hit him. The wind was strong and the cold was biting, even with his hat and coat. It was snowing heavily, and Dmitri was bundled in winter attire. They greeted each other, and before Dmitri could say anything further, the Professor grabbed his elbow and shouted through the wind, "Wait. Let me show you the rabbit skins out by the shed before you come inside." It was the excuse they needed in order for them to go outside and talk.

It was already very dark outside, with only the faint illumination of a porch light providing any relief. They walked through the thick snow, away from the house, for twenty or so yards until they reached the old tool shed. Then they stopped, and the Professor leaned over to talk directly into Dmitri's ear so that he could be heard through the howling wind. "It's been so long; still no access to the site?"

"No. You have to be patient, Professor," Dmitri responded, struggling to be heard. "If it happens at all, it will take time. You know as well as I do how slowly bureaucracies move."

As if to provide a visual emphasis of Dmitri's statement, the wind increased and blew snowflakes into their faces, making it more difficult for them to converse. Snowflakes draped each of them, and even became lodged in the Professor's beard.

"Yes, I know," the Professor answered, "but I feel like the claw is closing in on me even stronger by the day. The surveillance on me has increased like never before; I can feel it. It's all I can do to keep my sanity."

Dmitri leaned closer to the Professor and put a hand on his mentor's shoulder. It was courage and reassurance that he wanted to instill in him. "You need to give me the names of the others," he said. "I need to complete the arrangements. There's not going to be a lot of time if and when things get started."

Dmitri leaned back from talking into the Professor's ear and looked intently into his eyes. The wind howled and the snow flew between them as Dmitri watched the Professor consider the request.

The Professor had, in the past, alluded to the fact that there were other Russian scientists that were collaborating to produce material for their Chassis and Armor submission. Dmitri wanted to know how many of them there were, and he wanted their identities, so he could continue with the preparations for carrying out their plan.

Finally, the Professor leaned toward Dmitri's ear and said, "Including you and me, there are five — *The Mighty Five*, as I like to think of us." The Professor's referring to the group as "The Mighty Five" was a reflection of his devout affection for five famous, Russian composers of the mid-nineteenth century that split from the older European style of music to create music that was distinctly Russian. In a broader sense, it also symbolized his love and yearning for the Russia of old...for the greatness of it.

Dmitri wanted more; he wanted their names. The Professor knew this, and he was deeply concerned by Dmitri's request, regardless of how plausible it may have been. The Professor could barely reach out to any of those colleagues as it was because of the tight surveillance on him. Moreover, because they had not received any indication of their Chassis and Armor submission being accessed yet, he thought it was too early to risk disclosing their identities to anyone—even Dmitri. If and when the special website was accessed, then he would reveal those names. Not until then.

He told Dmitri that it was the agreement he had with the other members of The Mighty Five. Dmitri removed his hand from the Professor's shoulder, turned his head away, and thought for a moment, for he was anxious. It was a significant development for him to have finally learned the exact number of scientists involved, but he desperately wanted to know more; he wanted their identities.

The Professor could see that his protégé was troubled in his thoughts. "Come, Dmitri. Let's go inside, warm our bones, and have our cognac over chess."

Dmitri knew that it was ultimately the Professor's call. It was the Professor that thought of the whole idea in the first place, as well as the plan to actualize it. But the desire to know burned intensely within Dmitri.

5

Into the Weeds

About five weeks had passed since Angstrom began digging through the morass of technical documents and electronic files. The portion of his brain that needed to comprehend the complexities of the scientific and engineering principles embodied in the design and operation of a MAV had been awakened and was being exercised like it had not been in a long time; perhaps not since the days of his graduate work some twenty-five years ago. He was becoming well versed in chassis and armor technology, and he was even becoming knowledgeable with the entire MAV system. It was a test of stamina and concentration to review the extensive amount of information that his team had to go through, one after the other, day after day, week after week. The technology was fascinating, but reviewing it in such a forced, determined manner was essentially an overindulgence—scientific discourses, engineering figures, data tabulations, test and simulation results, mathematical computations, structural analysis—it was an immense endeavor. What made it slightly more tolerable was the unexpected sense of humor that Angstrom discovered in Keplar, and the fact that Angstrom and Keplar got along together so well. Hours were passed in listening to Keplar tell stories about his exploits in college—and of his many female conquests while he was there. Angstrom appreciated the direct, unpretentious way that Keplar expressed himself, even if sometimes it bordered on juvenile. If not for the great rapport between them, the days would have been much longer than they already seemed.

Angstrom had started the routine of coming in earlier in the morning and then leaving earlier in the afternoon. He needed the extra time at the end of the day to do something to reinvigorate himself after being locked up in the data room all day. A run on the treadmill, some weights lifted, anything to get his endorphins flowing.

As early as Angstrom would get to work—sometimes at the crack of dawn—Fred Book was always there earlier. It was automatic, the same time every morning; when Angstrom would approach the building, Book would already be there, standing outside in the cold, smoking a cigarette in the small area designated for smokers—Book, the "Pillar of Longevity." That was how Angstrom thought of him. Like the grooves in a vinyl record album, only endless without progression, circling over and over again on top of themselves, a rut more so than a guide, and the needle of the player caught in the rut and aimlessly moving along, with nothing produced for the effort. The needle had reached the end of the groove and was stuck against the ending.

Keplar had his own struggles with maintaining his energy and enthusiasm while enduring the routine of poring through submission after submission, but he trudged through it and was almost through with the task of initially filtering all of the submissions. He ruled out a great amount of material and physically placed those that were paper-copy submissions in a certain area of the room that they had agreed to designate as a place for the clear rejections.

It was Friday afternoon, the time of day at which Angstrom and many others would have left for the day, and both Keplar and

Angstrom were anxious to begin their weekends, but they were working a little longer because they could see the light at the end of the tunnel in terms of their first screening of the submissions. Angstrom leaned back in his chair and put his hands behind his head to stretch a little. The clock on the wall said 4 p.m., and he was about ready to call it a day, which is what he had just indicated to Keplar.

"Yeah, I'm about ready too, but I committed to myself that I'd get this pile over here done before the end of the week," Keplar said as he motioned to a relatively small pile of files and binders.

Angstrom looked at him with a smile, his hands still behind his head. "It'll be there for you on Monday. Let's get out of here."

Keplar stopped what he was doing, looked up at the clock, and then went back to work. "You go on ahead. I'll close us down. I'm going to make it through this pile if it kills me."

Angstrom laughed a little, leaned back even further in his chair, and looked up at the ceiling. In actuality, he had no plans for that evening, and given that it was the end of the work week, he all of a sudden became less in a hurry to leave; he felt calm and relaxed as he sat there, with Keplar toiling away next to him. He heard the sound of a binder being closed shut—another review of a submission complete, no doubt—and he closed his eyes for a moment to rest. A few minutes later he was startled when he heard Keplar laugh out loud, and he opened his eyes to see what the cause was.

Angstrom saw Keplar looking at a pornographic magazine. He watched him, and after turning through the pages, Keplar held the magazine outward and let the pages of the centerfold fall open. He whistled in an exaggerated fashion, then closed it up, tossed it to Angstrom along with its packaging, and said, "Here, sailor...make yourself useful. Consider it a gift for the weekend." Angstrom picked up the magazine—which had landed on his lap—with a puzzled, comical look on his face. "See what you think of that submission," Keplar said.

"You've got to be kidding. Someone sent this in as a submission?"

Keplar continued to laugh and said, "You got it. Chalk it up to our incoming mail inspectors: if something was deemed safe, they didn't filter anything out. As long as there were no traces of Anthrax, or anything else of the sort, it got forwarded to us. See what you think of *that* armor."

Angstrom shook his head in disbelief. "I'm gonna go get a cup of coffee. If we're going to be here for a while, I need some caffeine. You want some?"

"No thanks," Keplar said, still laughing at the absurdity of the magazine sent in as a submission. "You know, you really don't need to stay here just because I am. Really."

"Yeah, well...my life is pretty sad at this point...not much on the schedule, so I'll keep you company—at least for a *little* while longer."

Angstrom tossed the magazine onto his chair as he exited the room, and there it remained, along with the courier envelope in which it was sent.

Angstrom walked through the hallway until he reached the break area where the coffee vending machine was located. As he reached into his pocket to get some change, he heard voices coming from down the hall. For some reason, the sounds caught his attention, and he leaned sideways to look over and see who it was. It was a woman talking, striking in appearance. She had brown hair pulled up into a tight bun, a slim, very attractive figure, and was stylishly dressed: red-rimmed glasses (giving her a professional, almost academic look), a matching red, thin belt that sharply contrasted with a black, short-sleeved top and matching black skirt, and red, four-inch heels.

Her overall appearance was far more stylish than he was accustomed to seeing at the TTO. It was both professional and alluring at the same time. Even as far away as he was from her, he noticed her shapely and athletic figure, and when she smiled, which seemed to be often, her teeth gleamed brilliantly in contrast with her supple, red lips, which were themselves accentuated by a deep-red shade of glossy lipstick. It was not too often that a woman with four-inch heels walked through the halls of the TTO. She simultaneously exuded an aura of both professionalism and sexual force.

He put his change into the vending machine and continued to study her as the machine spit out a cup and brewed the coffee. At one point she looked over at him momentarily as she continued to talk with the other person. Angstrom could not get over how her formal, professional appearance seemed to be exaggerated and accentuated by her attractive and voluptuous overall look.

As he lifted the plastic door on the vending machine to grab his coffee, a colleague approached and stood ready to put his own change into the machine. "Chip, who's the woman down the hall? Have you seen her before?" Angstrom asked.

The man was startled at first because Angstrom rarely spoke to him. He rarely spoke to anyone around the office. Chip was surprised that Angstrom even knew who he was. "Uh, yeah, that's Susan Rand," he responded.

Angstrom was about to take a sip of the steaming hot coffee, but he stopped himself mid-stream and instead turned back again to look at the woman more closely. Now he knew why he thought there was something familiar about her; he had seen her picture before from the file that Seavers had given him. Even though it was over a month ago since he received the file that included her picture in it, he still remembered her face. She was supposed to be one of the members of his team, the as-of-yet unseen member, who up until that point still had not made an appearance. He had been wondering when he was finally going to meet her, because the impression he had gotten from Seavers was that he was going to be seeing her quite often and that she would be around a lot—instead of not at all, which was presently the case. Just as he made the decision to approach her and introduce himself, she broke from the person that she was talking to and began to walk toward him. He stood and watched as she came nearer.

Rand thought it was quite bold of the man to look at her as intently as he was. She was used to being gawked at given her attractive appearance, especially in such a male-dominated setting as DARPA, but she felt like the man was looking at her differently than just the normal sexual sizing up. He was studying her; his gaze had an unorthodox intensity to it. As she approached him, she had no inclination to acknowledge his presence, because she thought it was rude of him to be looking at her the way that he was, so she purposefully looked down the hall past him rather than at him as she walked.

"Susan Rand?" she heard, just as she was past him.

She stopped and turned back to look with a surprised, curious expression on her face. "Yes, may I help you?"

"I'm John Angstrom. We haven't been formerly introduced."

She stared at him while trying to fathom who the man might be; his name had still not registered with her.

"I'm leading one of the groups for the Chassis and Armor submission review," he continued, noting that she did not recognize his name.

After a short pause, she said, "Oh...yes...John Angstrom. I was told some new people were joining the effort." She looked at him with heightened interest. "It's a pleasure to meet you," she said as she held out her hand. "I was just on my way to your data room to sift through things; I didn't think anyone from the team would still be around so late on a Friday afternoon."

Angstrom did not appear anything like the man she had envisioned, now that she recalled Seavers informing her of a new Program Manager joining the effort. It was her turn to study *his* physical appearance. He was tall, mature looking, and handsome, but with an energy that seemed to be in check, ready to emerge at any moment. His dress was not too formal, but it was also not casual; just enough to look professional without the appearance of consciously worrying about it. Angstrom recognized the fact that she was taking note of him. Like herself, he was used to it by members of the opposite sex, and his overall serious and reserved demeanor usually presented an aura of mystery about him that heightened the reviewer's interest.

"Yes, Andersen and I are still here. I'll take you over there.

Would you like a cup of coffee?" he asked as he motioned to the machine.

"No, thank you. I can't drink the stuff that comes out of that machine."

"Yeah, well, I guess I've gotten used to it," he said as he looked at his cup with a smile.

"Lead the way," she said as she held out the palm of her hand to indicate the path down the hallway leading to the data room.

Angstrom swiped his security card and held the door open for her as they stepped into the room. He could see over the back of Rand's shoulder as she entered that Keplar was sitting in one of the chairs with his back to the door. His feet were propped up on the table, and he was paging through the adult magazine that Angstrom had thrown onto the chair. Keplar didn't move when he heard the door open, thinking that it was just Angstrom.

"Andersen, I'd like you to meet Ms. Susan Rand..."

Keplar paused. Angstrom saw his head dart up from the magazine and face forward with the realization that someone else had entered their area—a woman no less—and Keplar instinctively threw the magazine back onto the chair, like it was the last thing in the world he wanted to be caught holding. Then he stood up and turned to face Rand, his face already slightly red with embarrassment.

"Well, I can see you gentlemen are hard at work in here," she said as she laughed, turned from Keplar to eye the magazine still lying opened on the chair, and then back at Keplar again.

As soon as she took a closer look at him, she realized that she liked the way he looked, just like she did Angstrom, only Keplar's appeal was for different reasons as compared to Angstrom—most especially it was because of his age—his handsome, youthful appearance. Now there were two attractive men she had met that afternoon, and she felt like things were getting interesting very quickly. Rand decided that she would have to start making visits to this data room more often.

Angstrom laughed nervously and responded, "Yes, well, we turn it down a notch around here on Friday afternoons. Allow me to introduce Andersen Keplar."

Rand still had an animated look on her face, but Angstrom also saw something more. There was a peculiar level of interest he detected in her eyes as she looked at Keplar and shook his hand; her expression seemed to go beyond the normal smile associated with meeting a new colleague. She was looking intently at him, and Keplar, for his part, was drawn into her stare. Their handshake lasted a little bit longer than normal.

Angstrom decided that he would have to keep an eye on that; the memory of Keplar expressing the intention of picking up those ladies at the restaurant after their post-shooting range dinner came to mind. Rand was Angstrom's age, many years Keplar's senior, but Angstrom knew that one's age, or such an age difference, did not mean anything in the realm of sexual acquaintances. It certainly didn't in the case of his ex-wife's last fling.

"Actually," Angstrom finally interjected, "believe it or not, that magazine," he gestured down to it on the chair, "is an official submission."

She turned to look at him with a doubtful smile and said, "Oh? And I suppose Andersen was just checking for compliance with MIL-STD-499A, right?"

Keplar sheepishly laughed, but Angstrom ignored her comment and changed the subject. He turned serious and proceeded to explain all that they had been doing up until that point, and the process they had followed to filter out a first level of submissions. He motioned over to an area of the room that Keplar had designated for rejected submissions. Rand walked over to it and started to go through some of the material, and Keplar accompanied her, like a puppy following its master. He showed her some of the more ludicrous submissions, and they both laughed as he held up a model for their inspection. Angstrom went temporarily unnoticed by the two of them, almost as if he were a fifth wheel.

For some reason, Angstrom's eyes were drawn to the magazine that was sitting on the chair, and he saw that it was still opened.

A shiny Compact Disc encased in a clear, plastic sleeve was stapled into the binder between the opened pages. There was an image of a nude woman, crawling on all four of her limbs, impressed upon the surface of the CD. Something about it seemed odd to him. He didn't have time to focus on it, but it registered in his subconscious, and then he walked over to join Keplar and Rand in their discussion.

"Why don't we transition from the piles of material that are clearly rejects and look at some of the more serious submissions?" he suggested.

Rand and Keplar stopped what they were doing and looked over at him. After looking at her wristwatch, Rand said, "Thanks for the offer, but it's getting late. I've got to get to the airport to catch a flight back to Boston."

"Ah, alright. Well, perhaps we should talk soon, at least by phone, so I can update you on more of the details of our progress," Angstrom said.

"Or lack of it," Keplar interjected with a grin. Angstrom immediately shot him a glance, and Keplar got the message to keep his mouth shut from that point forward.

Rand noticed this, but ignored it, and instead she said, "You'll find that I'm not a micro-manager, John. You don't need to share with me every excruciating detail on the status of your efforts. Keep doing what you're doing, and when you're up to speed on chassis and armor technology, and have narrowed the submissions…say to around…six or seven…let's get together again and have your team share what you've found. Just keep the program schedule in mind, and have your preliminary results available early enough so that we can sufficiently analyze them in preparation for Phase-Two."

"Yes, certainly," Angstrom said with a slight appearance of puzzlement. "We'll keep that in mind," was all that he could force himself to say. It struck him as odd how she spoke in terms of her not wanting to micro-manage them, and how she presumed to give them instructions on how to proceed—as if *she* were actually the Program Manager and they were on *her* team. She was supposed to be a member of *his* team.

"You're both going to the pre-meeting, aren't you?" she asked.

"I'll be there. I'm not sure Keplar will," Angstrom said. He turned to Keplar and continued, "I have to check with Seavers to see what the cutoff level is for attendance."

Rand turned to Keplar and looked at him as she prepared to respond to what Angstrom had just said. She reached over and adjusted one of Keplar's buttoned down shirt collars. "He should be there; I would recommend it." She acted as though she already had a power over Keplar, control of him, and Keplar did nothing to dispel the notion. He just stood there and let her adjust his collar. Her making physical contact with him by presuming to adjust his collar did not go unnoticed by anyone in the room. It was almost as if there was something erotic about it, unspoken and left unacknowledged, but recognized by each of them nonetheless for what it was.

Angstrom tried to redirect the conversation. "I think it's important that we all stay on the same page with what we're doing so that we're coordinated in our efforts."

"Well, it's like I said. I don't normally get into the weeds at this point." She continued to eye Keplar. "You guys are on the right track; just keep doing what you're doing." She turned to look at Angstrom and continued, "And on that note, I've got to run, or I'm going to miss my flight. It was a pleasure meeting the both of you."

"Same here," Angstrom said.

A mischievous smile appeared on her face. She turned back to Keplar, wagged her index finger at him in an exaggerated fashion, and said, "Stay away from those magazines; they'll get you in trouble if you're not careful."

"Heh, right," he snickered.

"Have a pleasant weekend, gentlemen," she said before departing.

After she left, Angstrom and Keplar exchanged looks, almost as though to ask each other: What just happened?

It was Keplar who finally broke the silence and said, "Wow."

"Yeah, watch yourself with that one," Angstrom responded.

Keplar looked around the room at all of the material, categorized into different piles, and then he turned to Angstrom with an exaggerated smile on his face and said, "Well John, now I *have* had enough. Have a nice weekend; I'm outta' here."

They were familiar enough with each other by then that no further words needed to be spoken about the experience that they just had with Rand. Both of them knew that one aspect of their project had just taken quite an interesting turn.

"I'm right behind you," Angstrom said. "Let's go."

The door shut behind them as they left, but the lights remained on. They would remain so until the automatic system shut them off in another hour. The magazine lay open on the chair, leaving revealed the image of the naked woman on the CD, seemingly crawling on all fours through the ether.

6

Candy MAV

When Angstrom arrived at work early Monday morning, Book was standing outside in the cold, the same place as always, smoking a cigarette. The cigarette smoke could barely be distinguished from the steam that was coming out of his mouth as he exhaled into the cold, wintry air. It was not clear what progress he was making in his own efforts at reviewing the Chassis and Armor submissions, because he continued the practice of periodically coming into the data room, checking in one file, checking out another, and then leaving with the new file to go back to his office. He never bothered to report any of his thoughts or analysis about what he had reviewed, and Angstrom never made the effort to ask him. He was, in Angstrom's view, just plodding along.

There were not too many people moving about in the halls within DARPA's headquarters that early in the morning, and when Angstrom entered the data room, everything was as it had been left since last Friday. The cleaning people were not allowed into the data room during non-business hours, or when no one was in it, so everything remained untouched.

Meeting Rand on Friday had caused another piston to start firing in him; things were successively happening that were stimulating his consciousness—waking him from the slumber that he had been in for the last two years. Her hands-off, staying out of the weeds mentality did not sit well with him. He was a person of action and details, and to him, getting into the weeds meant getting things done. For all the work that he and Keplar had done over the past month and a half, she didn't show the least bit of interest or appreciation. He wondered whether she was going to be the type of person that took credit for someone else's work, unless something went wrong, in which case there would be enough distance between her and the issue that she would not get caught up in it. He was going to have to keep an eye on her and get a feeling for what she might be saying or doing, if anything, in conjunction with his and Keplar's work. Taking a hands-off approach with Book was one thing, but she was a different story.

He sat down in the data room and pulled the plastic lid off of his hot cup of coffee (he hated drinking coffee through a plastic lid). Steam rose from it, and as he raised the cup to his mouth to take a sip, he noticed the opened magazine still lying on the chair. For a moment he recalled how Rand had walked into the room and caught Keplar off guard looking at it, and he couldn't help but laugh.

But then his attention shifted to the CD visible from the opened pages, and he picked up the magazine to look at it more closely. The woman depicted on the CD certainly caught his attention. Her eyes were piercing and seemed to beckon the reader. Her hair was bleached blond, and her face was heavily made up—a thick dose of blue eye shadow, and an overabundance of a dark shade of lipstick. The face struck an earthy, almost animalistic countenance. It was a perfect example in the art of contorting the soft aspects of a woman's beauty so as to purposefully evoke a look of lasciviousness.

He pulled the plastic sleeve of the CD from its staples and continued to examine it. Something about it did not seem right.

There was no text on it at all; just the picture of the woman. The overall design imprinted on the surface of it gave it the appearance of having been made by an amateur on a home computer. A publisher would not have included such a CD without any copyright notice, loading instructions, or anything else whatsoever imprinted on the surface of it. He turned back to look at the cover of the magazine and then quickly flipped through its pages—there was no reference to the CD at all, which also struck him as odd. Why go through the trouble, and incur the extra expense, of inserting a CD into the magazine and not even advertise that fact?

What also stoked his curiosity was that publishers generally no longer even inserted CDs into magazines; it wasn't economically viable. It was barely feasible to print the magazines themselves, much less insert a CD into them. If there was additional content the publisher wished to offer, there would simply be a link provided that took the reader to an internet site, or even a bar-code for scanning which directed the reader to the site.

He looked at the image on the CD even closer, because he noticed a small black mark on the woman's inner thigh. Was it just minor ink spillage? A scar on her skin? He reached for a magnifying glass and held it over the area. There he saw, in extremely fine print and barely discernible, the words: *She isn't what she seems.*

An interesting teaser, he thought. But he considered why anyone would go through the trouble of including the text and then make it so hard to even notice, much less read. The average person would not have given the image of the woman herself anything more than a cursory glance at best before perfunctorily loading the CD into a computer, much less read such diminutive text.

Angstrom ruled out the notion that the phrase was a tattoo; he had seen some pretty strange ones before, but he couldn't believe that a woman would go through the trouble of having such a strange message inked on her body. Unless, he thought for a moment, might it be a brand? Some mark on her flesh, put there under the duress of the woman's pimp?

He was intrigued, and he wondered what could be on the CD itself. Advertising? More pornographic content? A feature profiling the woman? He took the CD out of its plastic sleeve and tapped it between his fingers a few times, wondering whether incoming inspections had scanned it for viruses. Concluding that it couldn't have because the seal on the sleeve hadn't even been broken, he decided to proceed anyway, and inserted it into one of the computer's drives.

A window appeared on the display, and a status bar indicated that the DoD's proprietary security software was analyzing the disc's content. When the analysis was complete, the software was supposed to list the contents of the disc, including information such as file type, file size, and last date of modification. Surprisingly, not much was listed for the CD: one JPEG image, and a single html link. No other information was presented in association with those two items, which puzzled him; the digital signatures for them had somehow been purged in a way that made it undetectable by the security software. Angstrom had never seen that before; it was a powerful security package and always found at least remnant traces of file information, and it reinforced his intuition that something was not right with the CD.

All of sudden, the computer, of its own volition, became active and appeared to be executing or further processing something on the CD. He wondered how that could be; the auto-play function was disabled on all DARPA computers, and the disc appeared to be formatted for standard file storage, yet something on it, some program, enabled it to take control of his computer. The security scan should have detected any phantom program capable of automatically invoking itself.

When the depression of a pre-defined sequence of keys on his keyboard failed to interrupt the process, he thought about immediately shutting down the computer as a last resort, but ultimately he decided to wait and see what might happen. A window opened in full-screen mode, thereby covering all of the display's icons and tool bars. The background on the display turned completely black, and the characteristics of the computer's underlying operating system were essentially hidden from view. A small, semi-transparent, green-colored square appeared in the lower right-hand corner of the screen and began blinking, seemingly indicating that a program was running, or installing, or something, he wasn't sure.

What happened next dumbfounded him: an enhanced, larger image of the same nude woman from the CD appeared on the entirety of the display. Given the context of where he had obtained the CD—from the inside of a pornographic magazine— the image should not have surprised him as much as it did, but the enlarged image of the woman, revealing her in full detail, disarmed him, and he studied it for quite some time. It was the exact same image, with one exception: the minuscule text that was impressed upon her inner thigh was replaced with a different set of text, which simply read: *Candy Mav.*

Because the text was different between the images, he reasoned that there must have been a specific purpose for it having been placed there; there was some significance to it. And again, it was done in a way for which the average reader would not have noticed the difference. He couldn't fathom what purpose it was meant to serve. And what was "Candy Mav"? The woman's stage name? But the last name—Mav—that couldn't have been a coincidence.

The green square was still blinking in the lower corner of the screen, superimposed now on top of the enlarged image. Angstrom grabbed the mouse and moved it around on the surface of the table, but it caused no movement to the cursor on the display. In fact, the cursor was gone. He clicked the buttons on the mouse, but it had no effect. He again tried hitting a certain key sequence to interrupt whatever the computer was doing, but try as he might, he couldn't call up the operating system's task management application. If it was not for the blinking green square in the corner of the screen, he would have thought that the computer was locked up.

He continued to wonder what was happening, and he started

to become concerned. Maybe the CD had a program on it that was erasing everything on the hard drive, or maybe it was instantiating a virus of some sort into the network. He was about to power down the computer when a black bar appeared at the bottom of the display. It was a blank text bar, and then an html link, written in hot pink font, faded into it:

http://www.imagecode/microsite/patternsubfile/index/ CLEV32/reflected.kz

He sat back in his chair and stared at the screen, intrigued. The country code for the link was Kazakhstan. He felt like the link was staring back at him, shouting at him, and compelling him to click on it.

He was not sure how or when it happened, but the computer's cursor reappeared on the display. Had it always been there, and somehow he just failed to notice it? No, that was impossible; he had tried to maneuver it before, and it wasn't there. Somehow the cursor had reappeared, all on its own, and now it felt like something was urging him to manipulate it... to move it over to the pink html link and click on it.

Reaching for the mouse and giving it a jerk, the cursor on the screen moved synchronously with the mouse's movement; control over the computer had returned. Just as he was about to move the cursor over to the link, the data room's door opened. Fred Book entered.

He stopped in his tracks when he saw the naked woman displayed over the entirety of Angstrom's computer screen. Angstrom, somewhat embarrassed—just like Keplar had been when he was caught by Rand looking at the magazine—realized that he had to say something, so he pointed to the screen and started to speak, but then stopped.

Book restarted his steps, held up his hand, and said, "Look, I don't know, and I don't want to know. I just need to extend my checkout of the last file."

Angstrom looked at him for a moment, bewildered because he was still mentally back with what he had just been doing, and also because he was not sure what to do about Book catching him in the present situation. Book had a cup of coffee in his hand and a file held under the same arm, and he just stood there, waiting. Finally it registered with Angstrom: Book wanted to extend the checkout of the file he was holding under his arm.

"Ah, alright. Let me see the asset tag number so I can look it up," Angstrom said as he held out his hand, expecting to receive the file.

"Just look my name up in the system; the file and asset number will automatically come up. It's the only one I have checked out, so it will show up at the top," Book responded with aggravation.

It was an awkward situation for Angstrom, because the image of the naked woman was still on full display. He looked back at it and was not sure what he wanted to do. It would have been embarrassing if it turned out that control of the computer was lost again, preventing him from being able to look up Book's information, so instead he turned the computer's power off, slid his chair over to another PC, and said, "Let's do it on this computer."

Book raised one of his eyebrows and said, "Fine, whatever."

When the work was complete, Angstrom turned to Book and said, "Alright. I've renewed the checkout."

"Grand," Book said dryly before he turned around and left.

Just as he walked out, Keplar arrived and made his way through the still opened doorway. He turned to watch Book leave, and a curious, playful expression appeared on Keplar's face when he turned back to look at Angstrom.

All Angstrom could do was gesture to him and say, "Don't ask."

* * *

Later that evening in his apartment, Angstrom had finished his dinner and was watching the BBC on television. There was turmoil in Syria. Even from as far away as the confines of his home in the United States, he could tell when certain events around the world were stimulated by his Government; field operatives, his old colleagues, injecting the will of the United States, and stimulating events in its favor.

After being out for two years, he was finally able to watch such reports with an emotional detachment. The doctors had told him that it would take a while, but that eventually such ability would return. He continued to be conscious of the healing that was occurring within him. The pieces of his life were being collected, re-assembled, and rewired, bit by bit, into some new semblance of him.

Eventually he got up and went to get his tattered, soft-leather briefcase and reached inside of it for something that he had been thinking about all day, and ever since he had arrived at his apartment: the magazine and the CD. It was against regulations for most people to bring such material home, especially without even logging it out in the department's online system. Not for Angstrom—not with the special security clearances that he still possessed.

Without hesitation, he inserted the CD into his PC and caused it to reproduce the same image that had appeared on the computer in the data room. He moved his mouse over the pink html link below the nude image and clicked on it.

After a few seconds of processing, the computer replaced the html link with a narrow text box, and a cursor flashed inside of it, waiting for input. At the very left-hand portion of the box was a simple message: *Enter Code*. Angstrom smiled and thought about how it was just as the text on the inside of the woman's thigh had indicated: Things were not as they seemed.

Before he did anything further, he went to the refrigerator, grabbed a bottled water, and looked out of the window of his twenty-first floor apartment into the darkness. Headlights from the traffic outside moved in every direction, like a lot of little fireflies buzzing around in a maze.

The computer prompt was asking for a "code."

After taking a drink while still looking out of the window, he went back to his PC and put the plastic bottle down next to the keyboard. If the CD was legitimate, there was only one plausible code that he could think of, and he carefully entered it into the computer: ZXX-10-09 MAV. It was the published TTO designation for the Chassis and Armor crowdsourcing project.

He knew that he was crossing a threshold: potentially top secret material exchanged over the internet—and from his home PC no less. He envisioned the code being transmitted to some remote server in Kazakhstan.

A message popped up in the lower corner of his screen; the internet security software on his PC was scanning an incoming transmission. He felt somewhat comforted by the fact that even though the unknown program from the CD had taken control of his PC, just like it had back in the data room, at least it allowed his security software to still operate.

From the LED flashing on his computer case, it was clear that there was activity occurring on his hard drive. He guessed that something was being downloaded onto it. That was perplexing, because something was being downloaded onto his hard drive without his security software preventing it, or asking for his consent ahead of time. Some sort of clandestine software on the CD was bypassing the capabilities of his commercial security software.

The image of the naked woman began to change on his display. The pixels began to separate themselves. It was oddly reminiscent of something that he had seen once before; it was back when he used to attend periodic technology briefings in his previous role. One of those briefings dealt with what was, at the time, considered to be an area of research that was receiving a lot of attention: a method of covert data transmission referred to as "Steganography." It was a way of secretly embedding an information payload into another file, or carrier, such as in a JPEG image. In essence, a JPEG image could serve as an "envelope" and carry secret, covert information embedded within it, undetectable by the naked eye. A special computer program was needed to process the resulting modified carrier and extract the covert data, or essentially, to open the envelope.

In the present situation, the non-secret JPEG image of the nude woman was modified to have concealed within it the secret data. It was not actually necessary to visually depict the extraction of the secret data by showing the depixelization of the modified JPEG image, but for Angstrom, who otherwise would not have known that the process was occurring, it served as an effective signpost. Unbeknownst to him, the custom program to cull that secret data from the JPEG image had been instantiated onto his PC, and it was now causing the data to be extracted. The program then re-formatted the extracted binary data into a human-readable form, and placed it into a newly created file.

When the process was complete, the nude image was completely obliterated from the display, its corresponding window closed, and normal operation was returned to his PC. There was one subtle difference: an icon representing a new PDF file appeared on his desktop. The legend below the icon, the name of the file, had the same ZXX-10-09 MAV designation.

He opened the file and sent it to the printer, and while it was printing he quickly cycled through its pages on his PC; it was a Chassis and Armor submission, and it appeared to be a legitimate one. His countless weeks of studying submissions had by then made him fairly adept at the MAV technology involved, and he was able to rapidly comprehend much of what was provided in the file. It was most certainly genuine; key terminology that he recognized was described in detail, supplemented with intricate images and figures.

The question was: Why would the file need to be submitted in such a clandestine fashion? He figured the answer to that question might be as interesting as the content itself. He picked up and inspected the courier envelope in which the magazine was sent, and there was no indication on it as to its place of origination. His mind began working: a trace would have to be placed on the delivery to learn of its routing, and the CD would need to be analyzed to investigate the programs resident on it, including the steganography programming.

And what of the internet site itself? Kazakhstan. Was it an intermediate gateway, or the end destination? Could the location of the end server even be determined? It was certainly beyond his current level of expertise; the last time he delved into the intricacies of network routing was twenty years ago, when such technology was in its infancy. Such an investigation was, of course, beyond the charter of the TTO.

Before he got ahead of himself, he decided that he should take a closer look at the submission itself to see what was really in it. It all looked plausible enough, but the proof was in the details.

* * *

"Andersen, take a look at this," Angstrom said toward the end of the next day.

"What's up," Keplar said as he rolled in his chair over to Angstrom's to see what he was looking at.

"Look at this chassis," Angstrom said as he held up a figure from the new file. It was a diagram of a cross section of a MAV, revealing different aspects and features of the proposed chassis. About thirty lines extended from different portions of it, each one with a label to identify and describe a certain technical feature in the diagram.

"It certainly meets some of the minimum requirements," Keplar said while studying the diagram. "Interesting. Some of the lines don't have any labels on them. What does that mean? More to come?" he said with a little laugh.

"Not sure. But take a closer look at the armor," Angstrom said as he turned to another page, which showed a partial view of the chassis with an armored shell overlaid on top of it.

"Well...it's got the v-hull construction. Hmm, what's *that* all about?" Keplar asked, pointing to a part of the diagram showing a cross-sectional view of the armor, which revealed what appeared to be embedded electronic circuitry with more labels stemming from it.

"It's sensor-enhanced armor—small transducers embedded within the armor to provide real-time structural analysis."

"You're kidding. Sounds cool."

"You don't recall any of the other submissions having that, do you?"

"No, I don't. Not even from the big contractors."

"And then look at this," Angstrom said, flipping to another page. "It's a different layer of the same cross section. Look at what it's proposing"

"Reactive armor..." Keplar read aloud from the caption below the figure. "Hmm...what's that?"

Angstrom pointed to different parts of the figure that were identified with labels, and then turned back and forth between the figure and supporting text on another page, his finger tracing the words for Keplar to read them.

"So whose submission is this, by the way? I don't remember seeing this one," Keplar said as he tilted his head sideways and grabbed a portion of the document to see its front cover.

Angstrom cautiously pulled the document away and said, "I got it out of the slush pile; I was looking for a laugh. Let me look at it a little more to see if I can make some sense out of it. There's no identification on the front of it to indicate who the submitter is, so maybe it fell out of another submission packet by accident."

He was not ready to tell Keplar how he had gotten the file; not yet. He also didn't share other technical aspects in the submission—aspects he had never seen before in the context of armored vehicle technology. The submission involved cutting-edge technology, of that Angstrom had become convinced. It was on par with some of the most top secret, advanced technology he had ever seen or heard of before. In his opinion, the clandestine submission far surpassed any of the other submissions in that data room.

The problem, though, was that the entire submission wasn't all there. Certain technical aspects were mentioned, but the supporting information was nowhere to be found. Some of the numbered pages were even missing; whole sections were seemingly redacted—almost like teasers to capture the reader's interest. The other problem was that it ended abruptly—it was unfinished. There wasn't even an identification of the submitter. At the end of it, however, there was a cryptic message:

> See more of Candy Mav—five days from original access:
> *http://www.bigboobs.kz.*
> PrC

* * *

Later that evening, he thought about the submission as he ate his dinner in his apartment. The message at the end of the file referenced something called 'PrC.' He was stumped as to what that might stand for or signify. His best guess: "People's Republic of China." But the 'r' was lower case. Maybe it was a typographical error.

His intention that evening was to repeat the whole sequence of steps that had produced the file on his PC and see if reproducing the process would yield the missing pages. He started from the beginning and inserted the CD back into his computer. When the image of the nude woman appeared, he clicked on the pink-lettered internet link at the bottom, just like he did before. This time, however, when his PC attempted to access the internet site associated with the link, a message appeared: *dns_server_failure.* It was an indication that the internet address did not exist. Thinking there was a mistake, he rebooted his PC and tried it again. The same thing happened.

The site in Kazakhstan had been taken down.

He sat back in his chair to think, and eventually something

occurred to him. He opened the PDF file that was saved to his desktop, and after scrolling to the bottom of it, he looked at the cryptic message that was at the end of the submission. Then he highlighted the internet address from the message, and cut and pasted it into the address bar of his browser: *www.bigboobs.kz*. It was a new link.

Sure enough, a web site came up, but he was not prepared for what he saw. Images of naked women were splashed all over his monitor. It was hardcore pornography. He looked at all of the images, trying not to focus too much on the women themselves, and searched for anything that might be related to the submission, but nothing stood out. In order to leave no stone unturned, he clicked on one of the images to see what might happen.

It caused another window to open, overlaid on top of the first one, with more images of naked women—very explicit images. All of a sudden, his security software displayed a window in the corner of the screen to indicate that it was blocking pop-up windows, and then it indicated that it was blocking successive malicious attacks on his PC. It was a malicious web site, and his security software was working overtime to block the attacks. He had seen enough, and became concerned that his security software would not be strong enough to withstand all of the attacks, so he moved his cursor to close the web site. Despite his attempts, however, it would not close, and instead, every time he clicked on a window to close it, another window appeared with more pornographic content. In desperation, he depressed the power button on his computer to manually shut it off. He took a deep breath after his computer had finally shut down.

He had a terrible feeling that his PC might have become infected with viruses, so he immediately turned it back on to see if it worked properly, and all seemed to be in order. Just to be safe, he initiated a full security scan to ensure that nothing nefarious had been pushed onto it.

He was confused about what had just happened. His access to the initial link had been cut off, with no further indication of how he might obtain the missing portions of the PrC submission. And when he went to the new link, all hell broke loose. If it was not for the fact that the PDF file itself seemed to represent legitimate, important disclosure of technology, he would have thought that the whole thing was a hoax, just to lead unsuspecting users to the malicious site.

As he further considered the situation, he realized something, and he referred back to the message at the end of the PDF file. The instructions were: "See more of Candy Mav—*five days from original access*." It was too early; he had to wait for four more days to pass. He wondered what the purpose might be in building a five-day delay into the process, but one thing was certain: a lot of thought—meticulous, careful thought—had gone into the process for establishing access to the PrC submission.

7

The Meticulous Conjuror

Sometimes the planning and the work is not the most difficult aspect of an endeavor; it's the waiting—waiting to see if the preparation, the implementation, and the bait, will land a catch—waiting, that time in between the effort and the result, the source of so much hope, frustration, doubt, anxiety, and perhaps, disenchantment. It can try those with even the firmest resolve.

When the Professor drove his old, run-down Volvo up the gravel driveway leading to the shed at the back of his house, the tires spun and threw off stones as they lost traction in the thick, densely packed snow. At first he was not sure if he would be able to make it all the way up the driveway, and when the car lost its momentum, he stopped for a minute, rolled down the window, and leaned his head out into the air. The sun was bright, and the cold, fresh air felt invigorating in comparison to the hot, stale air blown into his car by its heating system. Then he allowed the car to roll backwards several feet, onto a more level part of the driveway, so he could get a moving start.

The effort was sufficient, and he made it past the middle,

sloped portion of the driveway, which was the most difficult part to get over. The driveway curved to the right, and just before he entered the last portion of it that extended to the rear of the house, he approached the old Oak tree that had abutted his driveway for over thirty years. It was tall, thick, and mature, its bare branches reaching up into the sky like the scraggly fingers of a witch trying to obliterate the clouds with a swipe of its claw. The first sight of the old, rusty farm tool lodged in the trunk of the tree startled him; so much so that he almost swerved off of the driveway. It was the blade of a sickle that had been driven squarely into the tree, with its wooden handle sticking out and piercing the cold air as if in defiance of it. It was the signal—Dmitri's signal—that the waiting was over. Access to the web site had been detected, which meant that the package had made it to its intended destination, and its contents were perceived for what it was. He could not believe it... amongst the plethora of all of those submissions, of all of that muddle, someone had discovered their PrC submission; one chance in a million.

He stopped the car next to the base of the tree, placed both of his hands on the top of the steering wheel, and rested his forehead on top of his hands. He felt like he could have stayed there for hours; the wind howled through the open window, and the mixture of the cold air with that of the car's interior, in combination with the excitement from seeing the signal—that sickle in the tree—caused him to fall into a momentary state of meditation. His plans were set in motion. The submission had been discovered. The course of his life would change forever. It was just like he had told Dmitri: There is no turning back now.

The code that Angstrom had entered at the web site not only caused the nude image, along with the steganographic program, to be downloaded onto his computer; it also sent an indication to a remote server, operating in the bowels of a warehouse in Kazakhstan, that access to the site had occurred. From there, the message had been relayed to Dmitri, who in turn lodged the sickle into the tree. The staged release of the Professor's PrC submission had commenced.

More crumbs would now need to be dropped in order for the trail to be followed.

It was the safest way that the Professor could think of to submit the information, while at the same time not putting too much out there if appropriate contact was never made. Now that someone had extracted the submission, hidden as it was within the JPEG image of Candy Mav, Dmitri and the Professor could be sure that they had their audience. It was time for the person serving as the software coder to modify the pornographic website; it was time to present the next image of Candy Mav, with more information embedded within it.

As he brought his small bag of groceries into the house, the Professor turned at the door to look around outside one last time to see if anyone might have been watching him. He reminded himself of the fact that he mustn't act any differently than normal; nothing must appear out of the ordinary. Stopping in front of the tree like that and relishing the moment was a mistake; it could've called attention to the sickle. He could ill afford such errors.

There was a modicum of joy in his heart, or maybe not joy, but hope. He sat down on his kitchen chair and thought of *her*. It was *she* he was doing it for. He had to...there was no other choice anymore.

* * *

The five day window had passed; it was time for Angstrom to access the new web site. The wait for those five days seemed like an eternity to him. His PC's internet security definitions were up to date, and a full scan of his computer ahead of time revealed no viruses—not even any tracking cookies. He was ready.

When he entered the address, the web page came up; it was the same one as before. There were about thirty little boxes splashed across the screen, and each one had a sexually explicit image of a woman in it. To him, the images were grotesque; the women were debased beyond mention. They held no appeal to him whatsoever. He had been involved before in shutting down some major, illegal operations associated with such sites—fronts for prostitution, or worse yet, sexual slavery; it represented the underbelly of the world.

His eyes scanned each row for something out of the ordinary (if that could even apply to such a website). So far, no intrusive pop-up ads, or warnings of malicious attacks, appeared on his computer. As long as he did not *click* on anything, the web site seemed innocuous, but there was nothing there that he could see that was related to the PrC submission. After some hesitancy (he was concerned with triggering a PC attack), he moved his mouse to scroll down on the web page, revealing a whole additional set of images.

He leaned back in his chair, took a drink from the bottle of vitamin water resting by his keyboard, and stared at the display; not at anything in particular, just the muddle of misused and unholy flesh splashed across the screen. Banner ads running along the side of the screen were even more graphic and sexually explicit. Some of them had moving images of copulation. Then he noticed something; he put the bottled drink down, leaned over to grab the original CD with the nude image on it, and looked at it for a moment, and then back at the display. He wasn't sure, but one of the women seemed to resemble the image of the woman on the CD: pearl white skin, blond hair, thick black eyeliner, blue eye shadow, and dark lips. Her legs were spread wide open, leaving nothing to the imagination. Curiously, unlike the other boxes with pictures of naked women in them, this woman's image had no text or tag line associated with it—just the image of the woman herself. That was what initially caught his attention and caused the image to stand out from the others. A casual visitor of such a site wouldn't even have noticed the discrepancy. Certainly, he thought, no one else within the TTO. It was thus providential for the Professor that of all of the people within the TTO that could have received that magazine, it landed in the hands of someone like Angstrom.

The face in the photo held his attention. It was so heavily made up that he was not completely sure whether it was the same woman or not, but it looked like her, and it was all he had to go on, so he took a chance and clicked on the image. He was alert and at the ready in case the site tried to take over his computer again in a malicious fashion. To his relief, nothing drastic immediately happened—no pop up windows, no warnings of attacks, nothing.

Then something did occur: that same image of the woman on the CD, Candy Mav crawling on all fours, appeared in an enlarged image on the display. So there *was* a connection—it was her—he had clicked on an image of the same woman. She had that look in her eyes again, like she was speaking to him. His eyes moved across the image and traveled to her inner thigh, where he expected to see the text. There was something there, but it was different yet again.

It read: *Save me from the hell.*

Even with his extensive background in the field, and everything that he had experienced and witnessed over the years, a chill went through him, sitting alone as he was in the dark quiet of his apartment.

He stepped away from his PC for a minute because the message so affected him, and he went to the large window in his apartment to look out of it. He wanted to feel the cold of the outside, and he leaned his head forward until the skin of his forehead made contact with the cold glass. *Save me from the hell*— he went over the message in his mind again and again. Who was she, and what did she want to be saved from? Was she even real? Angstrom began to formulate a multitude of theories.

When he eventually sat down in front of his PC again, he noticed that a message had appeared at the bottom of the screen, below the image of Candy Mav, and similar to the way the text box had appeared for the last image. It read:

To see what's really behind Candy Mav,
enter your age code: _____

Candy Mav—linkage to the PrC submission was confirmed. The web site was prompting him to enter his age, just like any other adult web site. Even though it did not really matter what he entered, so he thought, he entered a fabricated age in case someone tried to trace the information back to his IP address and do something nefarious with it. Someone, or some organization, was clearly very savvy with coding and hacking, and he didn't want to take any more chances than he had to.

After entering the age thirty-seven, his heart sank. Pop-up windows began appearing at a furious rate, overloading his display with them. His security software feverishly displayed warnings of hacking attempts, one after the other: *intrusion attempt prevented …Trojan horse negated….*

Once again, control over his PC was lost and no matter what pre-determined key sequence he frantically entered, he couldn't shut down the web page, so he reached over and manually reset the computer.

He sat back in his chair to consider what had just happened. He couldn't understand why what happened actually did. It was the same woman, and her last name was given as Mav—obviously in reference to "MAV"—but when he entered an age, chaos ensued. He looked across the room, out into the blackness of the sky through the floor-to-ceiling window, and continued to reflect on what had happened when he entered the age.

Might it have known that the age was incorrect? Unlikely. Angstrom could not believe that someone would have known in advance that he would be the one to receive that magazine, and thus be the one to enter the site, and then know his age in advance. As he thought about it, there seemed to be a million possibilities, a few of them quite disturbing.

He was reluctant to try anything further because he didn't want to keep going through the hassle of rebooting his PC every time something didn't work. Then a more practical idea occurred to him. It seemed so obvious that he feared it couldn't be that simple. The prompt on the site was not asking for an age; it was asking for an age *code*, but there was no such thing as an age code, and he had a hunch as to what it meant: the prompt was asking for a code—*the* code.

Before he did anything else, he had his security software run an abbreviated scan to check for viruses, and as it did so he went to his kitchen and grabbed an apple out of the refrigerator. When the scan was complete, he was happy to see that, miraculously, his computer was not infected. The security report indicated that there were several attacks on his PC, but nothing had gotten through. He went to the site again, anxious for the web page to open.

When he got to the prompt asking for the age code, he entered: ZXX-10-09 MAV. He would've thought of it in the first place, but the pornographic nature of the site, combined with the reference to *age*, threw him off the trail.

The screen changed, but the site did not begin attacking his computer as it did before. Instead, an enlarged version of the new Candy Mav image took over the entirety of the screen. That was all that happened. He right-clicked on the image and saved it to his desktop, and then he tried left-clicking on it, but nothing happened. At least he had it saved to his desktop, and that was a good start.

He closed the internet, referred back to the desktop, and located the icon representing the new JPEG file. Double-clicking on it did nothing except to present the image back on the screen. Double-clicking on the opened image also did not cause anything to happen. It appeared to be a standard JPEG image.

He closed the file and this time right-clicked on it to bring up a menu listing the different types of programs on his computer that he could use to open the JPEG file. It was just as he had hoped: among the list of programs was a new one that had been pushed onto his PC. It was called: PrC_STEG_Extraction.

It seemed eerie how everything was essentially laid out for him. Someone had done a lot of advanced, thoughtful planning for everything to work as seamlessly as it did. It was meticulous work.

A small window opened to indicate that a steganographic extraction was in progress, and a status bar kept moving sideways until it was completely full, thereby indicating that the extraction was complete. When it was finished, a new PDF file appeared on the desktop labeled "MAV II." He could not help but smile.

Upon printing the document off, all two-hundred-fifty pages of it, he flipped through it and could see that the missing pages were there, and text inserted where there used to be blank spaces. There were also many new concepts added. The new information looked to be very complicated; more complex than anything he had seen before. Someone very smart was behind all of this magic; he was sure of it.

8

Dmitri

For as diligent as the Professor had been in his preparations, even he could not foresee every possible deviation that might occur in his plan. Many potential branches already existed in the decision tree analysis, and while he endeavored to account for as many as possible, he fully appreciated that trigger points existed that were out of his control: chance, human error, and so forth. The existence of just one additional deviation would cause the decision tree to expand significantly. Therefore, the fact that his plan had progressed as far as it had without a hitch was a testament to his brilliance.

With the Candy Mav site accessed, an important trigger had occurred, and a critical stage of the plan would follow. Destinies were now becoming firmly cemented for some, and for others, not destinies, but choices. Small ripples in the water were growing with momentum into waves, and events on both sides of the Atlantic were causing those waves to coalesce.

The phrase coined by the Professor (borrowed, to be more precise) for his cadre of co-conspirators was "The Mighty Five." It was the brain trust of the PrC submission. It could not have been created without them; not if it was going to be the cutting-edge vision of marine amphibious technology, and mechanized vehicular technology in general, that it was—a submission that would immediately garner the attention and interest of all who examined it. It would not have been possible for a single person, not even the Professor, to have the breadth and depth of knowledge necessary to assemble such a formidable document. It was, in a sense, a work of art—a masterwork— a figurative dance on the backs of those that schemed to constrain him—a treatise on the most cutting-edge military technology in existence. It was created so as to scream out to its audience: "Look at what this is. We alone present this to you. Note its brilliance, sophistication, and magnificence!"

It was treason.

Besides Dmitri Pavlovitch and the Professor, the remainder of The Mighty Five was alleged to have included three other scientists; friendships forged with the Professor over the last forty years—in essence a lifetime given everything that had happened to him. They knew of everything, and they were sad for him. They would help as they could. Not everyone, however, was in, or would be in, as deep as he was. They had their own lives to lead, families to worry about, and corresponding decisions to make.

The Mighty Five—in addition to serving as the name of the collection of scientists responsible for the submission, it represented his homage to certain masters of Russian classical music: Mily Balakirev, Alexander Borodin, César Ciu, Modest Mussorgsky, and Nikolai Rimsky-Korsakov. The name was chosen for particular effect, and the result of much thought and calculation, in order to have a manipulative, psychological influence upon one particular person: Dmitri. The identities of the three other scientists were known only by the Professor. That was how he wanted it, and it was agreed—agreed by everyone, that is, except for Dmitri. Dmitri thought that of all people, *he* should know their identities.

Now, because the Professor and Dmitri were at the next stage of their plan (a crucial stage), arrangements needed to be made, Dmitri insisted, and because it was *he* who would make those arrangements—that was his primary role in the whole plan, if truth were told—he should know the identities of the remaining scientists comprising The Mighty Five in order to complete the arrangements. He pressed the Professor to no end for their identities. Otherwise, he argued, how would he know what preparations would be necessary for them? How could he finalize their travel arrangements? It was already getting too far into the plan, which was what Dmitri said he feared would happen in the first place.

The yearning to know was like a malignant cancer for Dmitri, all-consuming with growth unabated, and that was just the effect that the Professor intended it to have. He knew his protégé well. The coined phrase, The Mighty Five, was deliberately chosen so as to feed that cancer; to stimulate it and cause it to fester within him; to consume him. The more the cancer was left untreated—by the Professor not revealing the remaining identities—the stronger the cancer's effect was on Dmitri. The Professor had full appreciation for the subtle complexities of human nature, and he used Dmitri's constant yearning to *know* to his advantage; it formed an ever-present distraction for the man, something that caused even him—the Professor's very own protégé—to overlook things and not see them when they were there to be seen. The yearning to know gnawed at Dmitri and clouded his judgment.

But it could not be put off any longer; certainly not without arousing dangerous levels of suspicion, even for one as distracted as Dmitri. So the cancer must be attended to; there would be a meeting. Rather, a series of meetings. The Professor would meet with each of the other scientists in series, he told Dmitri. He would update each of them as to the status of their plan and then seek their permission, once and for all, to disclose their identities to Dmitri. It was necessary, he would tell each of them, because the plans for the next stages needed to be made, and if they waited any longer, there might not be enough time. This was what he told Dmitri.

At first the Professor considered whether he should have indicated to Dmitri that he would meet with all of the other scientists at once. But what if Dmitri questioned such a proposal: "Why take such risk? Surely some of them must be from outside of St. Petersburg, and you've already said, many times before, that you're constantly being watched. Being observed at such a meeting with three other esteemed scientists would arouse too much suspicion and jeopardize the overall plan."

Was it likely that Dmitri would raise such an objection and actually consider the recklessness of such a proposal? Or, might he be blinded sufficiently by the cancer to not even consider it? The Professor didn't want to put the question to an actual test, and so he told Dmitri of his intention as he did. Individually and in series...in staged series just like the release of the PrC submission itself. That way all of the scientists could be updated and informed with the highest degree of secrecy and caution.

Dmitri liked the plan. He congratulated the Professor on his proposal, and he insisted that it must be done quickly. There was not much time. With all of the internet traffic, and the downloaded information...neither of them would want those traces to be out there in the *ether* for very long before the next phases of their plan were executed. The spymasters—the KGB— were vigilant and thorough, ever watchful, cocked and ready to move upon the slightest sign, Dmitri warned.

Yes, the Professor knew this, he responded, and he had already made the arrangements for the first meeting. The date, time, and place had already been set. Dmitri became giddy and almost salivated in anticipation of the disclosure—eyes widened, hands rubbed together in nervous excitement.

Calmly, the Professor disclosed the details of that first meeting. His deliberate pace in providing them almost drove Dmitri mad, but when it was complete, Dmitri was satisfied. The Professor was pleased as well, even surprised that no further questions were asked of him. Not the slightest hint of doubt or suspicion. Nothing like: "Why should we even have to go through the step of seeking individual permissions for the disclosures? Why couldn't you just tell me right then and there the first identity...and all of the identities? Couldn't you have sought that permission in advance? Perhaps made the disclosure of their identities automatic, triggered upon access to the PrC submission web site?"

Dmitri didn't think to pursue that line of questioning, and even if he did subconsciously, he didn't verbalize it. Perhaps it might occur to him later, after he and the Professor had already separated from their discussion by the shed, but not right there, on the spot. That yearning to know, finally fed, overtook his ability to effectively evaluate the circumstances. The cancer was satiated, at least for the moment, and it was doing its job superbly.

* * *

At the appointed time for the Professor's first meeting, Dmitri showed up after all, even though he wasn't supposed to. The Professor had suspected that there was about a fifty percent chance that he might, and there he was. It was just supposed to be the Professor and the first of the other three scientists, but there was Dmitri, entering the building, walking in like he didn't have a care in the world—like it was his right to be there—to be in on the secret. To be there at the *front-end* of things, like the Professor knew he always liked to be.

The Professor stood across the street, hidden from view, and clearly saw Dmitri enter the café. It was the first time that Dmitri had deviated from their plan and ignored protocol. It was brazen of him, but it wouldn't be the last time; the Professor knew that now. His *very worst* fear had not yet been realized, but one particular suspicion that he harbored was now confirmed: at some point, Dmitri was not going to follow their agreed upon plan as he was supposed to, and that point had finally arrived. Confirming his suspicion made the ruse that was occurring worthwhile for that knowledge alone (even though it concerned him greatly). He needed to know exactly where he stood with Dmitri. The plan called for absolute precision if it was going to be successful, including as much knowledge of all of the various facets associated with it as possible.

Dmitri, for his part, wanted to be there in person to see for himself with whom the Professor would meet. Then, even if for some reason the person still did not consent to disclosure, it wouldn't matter; Dmitri would have seen the person and discovered his identity. Besides, Dmitri figured, he already had his suspicions as to who the other scientists might be. After all, there were only so many scientists that had the necessary knowledge to contribute to their submission, and Dmitri knew all of the Professor's professional colleagues and friends.

But the revelations that day were not over just yet. Something else happened next that answered another of the questions in the Professor's mind—another of his suspicions, the biggest, most important one—and as a result, his worst fear was realized. The horrible truth had come sooner than he had expected. When he saw it, he questioned the very essence of human nature. Oh Dmitri, what could have been promised to you? What runs through your mind to make such betrayal worth it? What could cause you to turn against your mentor?

Even though the Professor had begun to have such a suspicion, it nevertheless stung him severely and shook him to the core when it was confirmed. Of course, the manner in which the answer was revealed—so suddenly, so overtly, right in front of his eyes—made it even more jarring. The Professor could see Dmitri through the café window, sitting at a table by himself, waiting for the Professor and the Professor's undeclared guest. When the Professor first saw what happened next, his fists tightened up, and his body tensed; a feeling of nausea rose up in him, and he became dizzy. He squeezed his eyelids shut and then opened them again, to make sure he was really seeing what he thought he saw—to make sure that it was not a horrid dream. A feeling of loneliness engulfed him; the man he took as his protégé, his closest friend, and who he all but raised as his own son, was now doing this to him. The Professor had hoped beyond all hope that his suspicion was wrong, but there it was: Beak-nose approached the cafe on foot! He was walking through the snow-covered sidewalk slowly, cautiously, heading straight toward the large window of the café that fronted the street.

The Professor forced himself to recover from the initial shock in order to watch the scene unfold with cold calculation; like he was looking through the scope of a rifle, straight at the head of Beak-nose. At first the man kept his distance and appeared cautious, being careful to not be seen by anyone, especially considering that it was still early and not yet the appointed time. Beak-nose kept a certain distance away from the café storefront, just as the Professor was, but to the side of the café rather than across from it like the Professor. The Professor also had the benefit of being shielded by the brick and mortar of the building that he was in, looking down through a window from its second floor.

Beak-nose stood at his post and monitored the people that entered the café. The situation continued in that fashion for about twenty-minutes: the Professor in his position, Beak-nose in his, and Dmitri sitting at the table by the window in the café.

When it was a half-hour past the appointed time, and neither the Professor nor the alleged scientist that he was supposed to meet made an appearance at the café, Beak-nose grew suspicious. He came to the conclusion that nothing was going to happen, and made a move. He walked right up to the large window of the café and stood there looking in. He grabbed a cigarette from the package in his coat pocket, lit it, and took a drag as he leaned his face closer to the window to peer inside. He knew the Professor was not in there, but he looked anyway. He saw Dmitri sitting there alone at one of the tables, right in front of him on the other side of the window, like some fool who had been stood up for a date. Dmitri saw him, became embarrassed, and shrugged his shoulders, indicating puzzlement at what had just happened, or rather, what had *not* happened.

Beak-nose wasn't puzzled. He was angry. He knew it was not

by accident that no one showed up, and he knew that they were exposed. He casually turned around and surveyed his surroundings while he continued to smoke his cigarette, trying to see if he could spot anyone that might be watching him, but he could see no one. There was someone out there, though; he could feel it, and he was furious.

9

The Secret of Man's Being

It took a long time to accomplish, but Angstrom read through and studied the entire PrC submission. He consumed it and exhaustively considered every detail of it. Other submissions were cross-referenced for comparison purposes, but more so to serve as additional background material to fill in the gaps of his knowledge. Even that was not enough; the submission covered too many concepts that were unfamiliar to him, and at a level of detail that was beyond him. He searched online for supplementary information on certain subjects, but that was of limited use. He would have needed graduate degrees in several specialized technical areas in order to fully grasp what was being disclosed. The submission was technically dense—page upon page of complex, mathematical computations in support of the technology proposed.

After he had completed his review, and read much of the content several times over, there were hand-written notes all over the printed pages of the submission, with arrows and lines attached to certain text and figures that extended outward into the margins where there was more room to write what he needed to, and citations to different extrinsic material that facilitated his comprehension, and even material that he had gathered himself and inserted in between the pages for quick access. Angstrom was convinced that not only did he have a "winning" submission, he had in his possession technology that was far beyond any of the other submissions. It represented an advance in the state of the art of many key aspects of MAV technology: semiconductor devices, signal and data processing, material science, electromechanical elements, and more.

He eventually arrived at the notion that it could not possibly have been the work of any one person; it had to be the result of a highly sophisticated team of scientists and engineers. He also became convinced that the submission had not originated from within the United States. There was something about it that did not seem quite right; the technology was so different than all of the other submissions, and even though it was written in English, he noted patterns in the phraseology that would seem to have originated in some other part of the world.

The question, then, was where *did* it come from? At first he thought about Kazakhstan because of the html link, but ultimately he ruled that out; the notion seemed absurd.

He had two other clues.

One was the apparent signature at the end of the submission: P-r-C. It also appeared at the end of the prompt that directed him to go to the "bigboobs.kz" web site to obtain the rest of the submission, though he had not attached any significance to it at the time. P-r-C: the "People's Republic of China?" he hypothesized. A team of scientists from China? Various scenarios ran through his mind. If that was the case, he wondered why it would be so. Why would China submit some of its most important, cutting edge technology to the United States Department of Defense?

He thought about it continuously over the course of the next week.

One theory he held for a while was that it was a rogue attempt at black market arms trade, or in this case, technology trade. But eventually he ruled that out; it was unlikely that such detailed and sophisticated technology would be in the hands of black marketers; they wouldn't have known what to do with it. And he couldn't see them having such sophisticated means for covertly delivering the PrC submission via steganography.

He came to believe that the most plausible theory was that the submission represented a proposal, a request, for the Chinese scientists to defect to the United States; the "ticket" offered was the technology, along with an implied promise of more of the same when the scientists made it into the country. It was the best theory he had, and the one that he held onto, even after considering his next clue, which was even more perplexing. It was completely out of context with the rest of the PrC submission. Right before the PrC signature, at the end of the document, a phrase was inserted:

For the secret of man's being is not only to live but to have something to live for.

When he had first read the phrase, it made him pause, resonating as it did with what he himself had been struggling with internally over the last couple of years: to find meaning and passion in his life, after all that he had gone through, and everything that was lost. The phrase kept repeating itself in his mind for several days after he initially read it. It affected him, haunted him, just like the Candy Mav image did (with those mysterious eyes penetrating him). It was as if the originator of the PrC submission, the lead author of it, the director, whoever it was, knew just what to do, collectively, to pull the reader in, to grab hold of his very psyche and not let go of it. When Angstrom thought about it over those several days, he even wondered again at whether someone knew about him and his personal circumstances, and whether the submission was targeted and geared especially for him.

Try as he might, he could not make that phrase fit in with the rest of the submission, for the submission itself was all about technology—MAV technology, math, physics, and engineering—and then there was that phrase at the end of it. He couldn't reconcile it with the main body of the submission, but he knew it was there for a reason.

Finally, he decided to take it at face value, and he attributed it, like everything else in the submission, to the Chinese scientists. His postulate: they had nothing to live for, no meaning in life, and as a result, they wanted out. Angstrom could envision it; he knew from first-hand experience the type of oppression that existed in China. He had been there and seen it. For Chinese scientists it was probably worse; hand selected at an early age after being identified as intellectually gifted, they would have no choice in life, forced to use their talents at the bidding of the Chinese Government.

Thus, those two clues in combination—the PrC signature, and the phrase at the end—were interpreted to be a sort of valediction which, inherently, identified the submission's creators. It was a theory, but *only* a theory; a vagueness.

* * *

It was the first day of the Phase-Two meetings, a kickoff to prepare for the next phase of the crowdsourcing program. Keplar was excited. Wanting as he did to pursue a career in Administration (Angstrom could still not get over that), he was keen on being able to attend a meeting that would have so many senior managers present. Keplar also didn't hide the fact that he was looking forward to seeing Rand again, and Angstrom made a mental note of his enthusiasm, principally on that point.

Rubbing elbows with management was the last thing on Angstrom's mind; he was interested in the meeting for a different reason. He was curious to see the caliber of people that were running the massive crowdsourced project, and he wanted to present some of the more intriguing concepts from the PrC submission. Because of the defection issue lurking in the background, he didn't want to propose the whole submission as a candidate; not yet. Too many questions would be asked, and he was not prepared to share the origin of it with such a large audience. In fact, he was not quite sure *what* he wanted to do about that. But he did want to present some of the technology to the panel of experts, have it be discussed, and gauge their reaction to it.

When Angstrom and Keplar entered the large conference room, it was already filled with people. The first row of chairs around the long conference table was surrounded by another row along the walls. Almost all of the chairs were taken, and several people were standing at the head of the table. It was a packed room. Angstrom had purposefully arrived right at the scheduled start of the meeting so that he and Keplar could grab their seats and not have to engage in small talk with anyone.

Seavers opened the meeting to kick things off, and spoke at great length. For the next several hours, others did the same, some from the TTO, others from the military itself, in full military dress.

It was a long day. While the technology was interesting, it was buried in so much administrative detail, project lingo, and jargon that Angstrom wondered whether thirty-percent of what was discussed was meaningful content. He looked at Keplar at one point, whose eyes were watering and hair was slightly disheveled, and he laughed to himself, thinking: Welcome to the world of Administration, Andersen.

That first day was relatively uneventful. None of the other Chassis and Armor teams had yet presented, and Angstrom was disappointed in the slow, deliberate pace of the meeting. It turned out to not be that interesting to Keplar either, for many of the same reasons, but also because Rand was not in attendance. It was a major letdown for him, and Angstrom bristled when Keplar revealed as much to him. The second day held some interest because the parallel Chassis and Armor teams began their presentations on the "finalists" they had selected. There were no surprises as far as Angstrom was concerned: major defense companies.

It was not until the third day of the meetings, however, that things really became eventful. The Chassis and Armor candidates would continue to be presented by each of the teams, and Angstrom's presentation, representing his team's efforts, would be the last of the day. The dynamic of the meeting changed on that third day, however, mainly for another significant reason: Susan Rand finally made her appearance. She had flown into Washington the night before, and when she entered the room the morning of that third day, it was a major disruption, and the energy level in the room immediately increased. She entered like a rock star, and everyone stopped what they were doing and got up to have a chance at shaking her hand and greeting her.

"Gentlemen, first of all let me extend an apology for arriving so late," she began when it was her turn to present. "It seems that even with all of our most modern technology, we still can't build a commercial aircraft that is immune to a little snow on the ground"—a slight chuckle around the room. "Anyway, let's talk about a fast, adaptable, next-generation Marine Amphibious Vehicle. I'm very pleased to be a part of this ground-breaking project...."

She presented slide after slide, but the men in the room weren't paying any attention to them; they were watching her, and her legs, and the back of her ass as she faced the slides on the screen and pointed to the details in them. She knew they were watching her and not the slides; that was why she went through the presentation the way that she did. It was an act; a performance for the benefit of the men in the room. The outfit she wore was chosen most particularly with the effect she intended it to have on her audience. Angstrom could smell her perfume even from as far down the table as he was.

"The old mil-standard 499A, the Systems Engineering Process, is being turned on its head. That standard discouraged modularity. It discouraged manufacturability. It's the old way of doing things in a serial process: system layout...subsystem layout...component design, and then re-design and re-development because the design did not take into account the interaction of the modules in the first place; the interfaces broke down...."

She went on and on, and every man in the room loved it. They could have listened to her all day. Angstrom studied her. Not like the other men who were drooling over her and undressing her in their minds. He was observing her mannerisms, detecting inflections and subtle accents in the different parts of her speech. He noted certain physical movements of her body when she expressed certain emotions. He was learning all about her from things she didn't intend to communicate, and he was filing it all within his memory banks. For Rand, the meeting was an opportunity to impress the men in the room and keep them in her hip pocket. For Angstrom, she was up on a stage for his careful study and consideration.

"Our goal, then, gentlemen," she continued, "is to shorten the development time and shift to a high-value design focus. We want to "Democratize" design through crowdsourcing, and *vehicleforge.mil* is the groundbreaking mechanism that will allow us to do just that. It's an open-source, collaborative development environment to support the creation of a complex system by unaffiliated designers. There's going to be custom collaboration on sites linking universities, think tanks, and defense contractors, both large and small. Collaboration sites to allow for open source version control, software development...."

Angstrom turned to look at Keplar, who at that moment looked as though he was melting in his seat. He was under her spell; they all were—everyone in the room. Rand had them wrapped around her finger; they were like little boys to her, putty in her hands. She was the highlight of the day—perhaps of the three days, judging by the energy in the room. Her presentation was right before lunch, and after lunch, on that third day of meetings, it was finally Angstrom's turn to present. Half of the attendees had already stepped out of the room during the break and had not bothered to return. Others were half asleep after three days of presentations and a full lunch in their bellies. He was not left with much time, either. Too many presenters before him had run over their allotted times, and the cumulative effect was that it cut into his time slot.

It did not matter to Angstrom. His focus was sharp, as it always was in such meetings, especially when he was presenting. He held a commanding presence—his physical size and attributes, tone in his voice, mannerisms, all captured the attention of those in the room. Some were just curious to hear him speak, after having seen him but never spoken to him for the last two years.

He also knew how to read the political tides, and he had prepared the majority of his presentation based upon his earlier conversation with Rand herself, such that his group's three finalists were all major defense contractors. Nothing controversial, and the content of the three submissions, in fact, looked very much alike; they were not even too different from the current generation MAV; just iterations of it. Everyone in the room seemed pleased, most especially Rand, and she said so.

Seavers verbally congratulated Angstrom on a job well done as well. He could see the presentations for what they were, but given his concern and skepticism about the whole program in the first place, he was satisfied with the outcome of it so far. He was also pleased with the personal growth that he could see in Angstrom. Seavers told himself that his instincts had been correct—giving Angstrom the assignment was doing him a world of good.

The completion of Angstrom's presentation marked the end of the meetings, and as everyone started to get up to adjourn, Angstrom cleared his throat in a manner so as to get everyone's attention again, and then he said, "Although we've presented our top three candidates from the group, I would like to turn your attention to a few other, additional aspects."

Seavers, who had just risen from his chair, was the first to sit back down, followed by everyone else in the room, with curious expressions on their faces, and in some, looks of aggravation, as they were very much looking forward to the end of the day.

"There were a lot of strange submissions we received for this program, as evidenced by the jokes we've seen in the previous presentations over the last several days," Angstrom said. "But there were also some intriguing aspects in some of the lesser-known submissions, and I would like to talk about that for just a little bit."

His plan was to highlight a few of the more fascinating aspects of the PrC submission, and then gauge the response of his audience. It would be a sanity check of sorts.

"For example, here's an extract of one submission that deals with the concept of sensor-enhanced armor." Angstrom showed a slide regarding the technology, with text on one side of it, and a cutaway view of the armor to show some of the electronic details on the other side.

"Whose submission did that come from, John?" Seavers interrupted right away.

"Well, I have to apologize, because I didn't track the sources on these last few slides. We've been going through so many submissions, I can't recall at the top of my head. At this point, I just wanted to make sure that we all at least had a chance to see some of these concepts, and if after that we all believe that it's worth it, I can follow up on that."

"I see. That's fine; as you say, we can come back to it later if there's something there," Seavers said. "After all, that's what *vehicleforge.mil* is all about: collaborative, real-time design, right?"

"Exactly," Angstrom responded.

Upon Seavers engaging in such dialogue, Rand stopped the side conversation that she was having and began to listen to what was being discussed more attentively.

"So here you can see," Angstrom continued, referring back to the slide, "the results of a simulation of a shockwave impacting the metal armor. This plot," he pointed to the slide, "shows the frequency spectrum of the reflected sound wave as detected by the embedded electronics. The reflection is converted into an electrical voltage that the system controller utilizes to determine, from the spectrum of the reflection, whether the armor plating was cracked as a result of the hit. Essentially, damaged plates will reflect a different spectrum than undamaged plates."

"That's way beyond the scope of any current manufacturing capabilities," Rand suddenly interjected. "Semiconductor technology has not caught up to the requirements necessary for the small transducers to survive the extreme temperatures, radiation, and quite frankly, mechanical shock that would be experienced in a marine amphibious vehicle."

"That may be true for current semiconductor technology," Angstrom calmly responded, "but this submission went further, and delved into the actual semiconductor technology itself that would be suitable for such a system. It proposes a novel Indium-Phosphide, heterojunction transistor for the sensor, and goes on to describe...."

Before he could finish his sentence, Rand interrupted again and said, "Indium-Phosphide has never been proposed for this type of operating environment. I would suggest that you stay focused on more realistic technology; otherwise, I'm afraid your team will get side-tracked on things that just aren't technically feasible to manufacture." She turned to look at Seavers while continuing to address Angstrom, as if to convey to Seavers what she was actually saying to Angstrom. "I mean, you organize your team how you want, John, I'm just trying to help you given how new you are to this effort."

Angstrom looked over at Seavers for a moment, and could see that Seavers was interested in the technology, but that perhaps for purposes of the meeting, he should just move on, so Angstrom progressed to the next slide. "Another interesting concept we found was for something referred to as 'reactive armor.' This involves special tiles that significantly improve protection against rocket-propelled grenades. These tiles," he continued as he pointed to a cross-sectional view of the armor on the slide, "makes the armor reactive in the sense that each of them has an explosive that is detonated only when hit by a high impact rocket. The detonation in the armor is designed to react to only large impacts, as opposed to the lighter impact that results from smaller caliber munitions or shrapnel. When the explosive of a tile erupts, it disrupts and deflects the impact from the incoming RPG, thereby lessening the effects of the hit."

"Yes John, we're well familiar with the concept of reactive-armor. It has some utility in certain applications, but obviously the drawback is that once a tile has exploded, it's gone and provides no further protection, so another direct hit in the same spot would not be protected," Rand said in an aggravated tone. She was miffed because she felt like the attention from those in the room had shifted from her and over to the things that Angstrom was presenting.

"That may be true," Angstrom said in response, "but here's the really novel aspect of the submitted concept. It combines the *reactive* armor with the concept of *active* armor. It's active in the sense that the reactive armor can be rejuvenated after a direct hit. A second layer of explosive tile is coupled to the first layer, and its explosive content is rendered passive until the first tile has been detonated. After the first tile has reacted, the embedded electronics causes the second layer to become active; it's sort of like applying a detonator to C-4."

Everyone in the room, including Rand, was silent. They had never heard of something like that before, and it intrigued them. Seavers was clearly intrigued. Oddly, though, no one said anything for quite some time, until finally, it was Rand who spoke and said, "It would never work. The explosion from the detonation of the first part of the tile would disrupt the second; there's no way to isolate the second tile from the detonation of the first to prevent it from detonating at the same time."

After Rand said that, everyone in the room turned to look at Angstrom to see how he would respond. He was prepared, though; or rather, the PrC submission was prepared. "Yes, well, the submission I was looking at considered that issue. Look," he continued as he transitioned to the next slide. "In addition to the reactive and active elements, the explosive tiles are made from a new composite that is capable of isolating the first blast from the secondary layer of the composite tile." He pointed to a cross-sectional view of the multi-layered tile, and then referred to a complex Lewis Structure diagram describing the organic chemical structure of the new material.

Now all heads turned back to Rand to see what she would say in response. The problem for her, however, was that she did not have the scientific background to comprehend the chemical structure; it was way beyond her technical level of expertise. It was also a matter of first impression for her. So, as a defensive mechanism, she spoke in the abstract and brought the conversation to more general terms, indicating as she did before, when they talked about the armor sensors, that science had not caught up to such concepts, and it would therefore be impossible to develop and manufacture it within budget. It would therefore go against one of the goals of the project: to design an economically feasible next-generation vehicle. She spoke for some time, quite passionately, and then skillfully moved the conversation away from Angstrom's presentation and back to the general manufacturing principals for the project as a whole. She in fact called a conclusion to the meeting before Angstrom had finished presenting the rest of his information, and no one seemed to notice or care—not even Keplar. All of the men rose and hovered around Rand as they firmed up their plans for attending the cocktail party that was going to be at a local restaurant immediately following the close of the meeting.

In the meantime, Seavers approached Angstrom at the head of the table, put his hand on Angstrom's elbow, and leaned over to say something quietly to him. Then Seavers left; he would not be attending the party. Rand saw the whole thing take place while she entertained conversations with three other men at the other end of the room.

Angstrom was the only man in the room not groveling over her; he stood there at the head of the table by himself and gathered his presentation materials. It bothered him how Rand had pushed back on all of the new concepts he presented, and he wondered if it was sheer vanity, or whether it was something else. His mind worked quickly, and he ran through various possibilities. The one that came foremost to his attention was the possibility that there was another entity that had her on retainer, and there was a conflict of interest for her. He turned his head up to look at her, and their eyes met.

The cocktail gathering was more of the same in that Rand was the center of attention. No one was really interested in anything that anybody had to say until they had their own personal audience with her. It would have been amusing to Angstrom if it wasn't so pathetic. One man after another stood with Angstrom and feigned interest in what he had to say, only to constantly look over at Rand to see what she was doing, and whether she might be looking back at them. It was juvenile, but he stayed at the party and trudged through it, telling himself that he had to spend at least a minimum amount of time there to show that he was a team player.

About an hour into it, Angstrom was by himself at a far end of the room, nursing his drink. As he surveyed the people, he noticed Keplar talking with Rand on the other side of the room, and he watched them with interest. He could have sworn that at one point he saw Rand reach over and stroke the side of Keplar's cheek. It seemed an odd occurrence. Did he imagine it? It happened so suddenly, and there were many people between him and the two of them; he wasn't sure. It would have represented another bold advance on her part—more so than a simple adjustment of a shirt collar. He tried to look around some people to get a better view, but by that time they had already separated. Angstrom made a mental note of the incident—another interesting development, if in fact it had actually happened. He looked down at his glass and saw that it was empty, and he decided to get another drink.

After a couple of hours, Angstrom decided that he had

adequately served his time, and he made his way toward the front lobby to leave. Rand noticed from across the room. She actually had her eye on him the whole evening, but she did her best to not lead on that she was interested in talking with him, waiting to see if he might come over to speak with her. When she spotted him leaving, she went straight over to head him off. Even when she met him at the coat check, however, she didn't say anything or acknowledge his presence; she provided the ticket for her coat with an intentional blank expression on her face, hoping that he would notice her and say something.

She was about to put on her coat when Angstrom grabbed part of it and said, "Allow me to help you, Susan."

"Oh, you're leaving too?"

"Yes. Early to bed, early to rise... you know how that goes."

As soon as they stepped outside into the cold, she grabbed his arm and said, "Come with me."

"Why? Where are you going?"

"It's too early for me to call it a night, and being from out of town, the last thing I want to do is hang out alone at a bar or something."

"Alright, so where do you want to go," he responded, still standing there with her hand on his elbow.

"Let's go someplace quiet where we can talk more about those concepts you brought up at the meeting."

"Really?" he said, with one of his eyebrows turned upward in a querying fashion, doubtful of her expressed intent. After a moment of considering the offer, and looking at the smile on her face, he said, "Okay, sure, let's go."

They walked through the parking lot together, avoiding the puddles on the ground from the melting snow, and before he knew it, and without too many words spoken between them, he was in her car. Eventually, she pulled it into the parking lot of her executive suite.

"We're going to *your* place?"

"Well, it's someplace quiet...." After he feigned hesitation, she teasingly said, "Oh come on, you're not afraid of a little ole' government consultant, are you."

He was not naïve and had been in such situations countless times, so he knew the implications. He played along, curious to see how far she would go.

There was nothing unique or personal about her place; it came furnished with adequate but rather sterile décor. They spoke to each other generally for quite some time, neither of them revealing too much about themselves.

They had a slow glass of wine together as they talked, and Rand poured them another after they finally finished the first. They were sitting on the couch, her body slightly turned toward his as they talked.

"So what was all that about today?" she finally said to him.

"What do you mean?"

"All of that stuff about sensory armor, reactive armor, Indium Phosphide...what *was* that?"

A confused look appeared on Angstrom's face. "It's just what I said. Splinters of technology from some of the other submissions, things that I thought might be useful to consider along with the final candidates."

"Yeah, well, don't catch me off guard like that anymore. I'm not used to it." She leaned over and put her glass of wine on the coffee table in front of them, and then put her elbow on the back of the couch and rested her head in her hand as she looked at him and said, "Run things by me first, okay? I've done this a lot, and I know the direction those men like to go." He didn't say anything in response, and she studied his face as he looked at her in silence. "Stay in the lane with me and you'll be very successful, along with the project itself."

"So you don't want to consider some of those other aspects, or at least have some of our experts take a deeper dive into them?"

"Look, it's just going to be a distraction, and quite frankly, a lot more work for everyone. Stick with the tried and true and we'll be fine. You know what I mean?"

"Yes, I think I see what you're saying."

Rand could tell that Angstrom was not convinced and didn't

agree with what she said. She tried to get him to drink more wine, but the faster she drank hers, the slower he seemed to drink his. She couldn't read him—couldn't tell whether she should pursue him or not, but for many reasons, she wanted to. He was certainly handsome, but she had been with handsome men before. In fact, she had received many proposals in her day from good looking men in powerful positions, but she had always turned them down. Angstrom was a unique combination to her in terms of a conquest. He passed the first threshold of being good looking. Beyond that, she could not quite put her finger on why she was so drawn to him. She wanted to control him, like she controlled all of the other men she worked with in the TTO (except, that is, for Seavers). That was, in actuality, a big satisfaction that she got out of her consulting engagement with the TTO. It wasn't the technical aspects of the subject matter involved—she could get that at a lot of places—it was the challenge of trying to control all of those men—to get them to feed out of the palm of her hand. She focused just as much on that, or even more so, than the work itself.

For the most part it was easy. She didn't even have to sleep with most of them. Flirtation was enough. It was a delicate balance, doing just enough without going too far. On occasion she might run into a situation where she would have to go further, especially when she needed to wield a certain amount of power, but not too often. Many of them were conservative, happily married men just satisfied with a little tease now and then. Angstrom disrupted all of that; if she wasn't careful, he would be the antidote that would break her spell over all of those men. The danger revealed itself when he started presenting all of those new concepts at the meeting, and everyone in the room became intrigued with them. And then, at the cocktail hour, he didn't even show the slightest interest in her.

She was getting antsy that evening at how slow things were moving; usually someone would have tried to make a move on her by then, or at least sent her some kind of an indication. After all, she reasoned, he did agree to come to her place.

Angstrom watched her as she sat in silence, pondering her thoughts. All of a sudden she took the empty glass of wine out of his hand and put it on the coffee table next to hers. He watched her, curious as to what would happen next. He was pretty sure he knew. She gazed into his eyes, both to read him and to attract him. Then she slid closer to him on the couch, and even though he still didn't say anything, or respond one way or the other, she thought she sensed an opening, a willingness in his eyes. She leaned over and kissed him. When he didn't resist, her kiss lasted even longer. She had a serious look on her face when they separated, and then she stood up and said, "Wait here, I need to get something."

He tried to imagine what she would look like when she came back. He himself was still in full control of his faculties and emotions.

When she returned, she was dressed the same as before, but there was one subtle difference. As she came closer to him, he could not believe what he saw; she had a leather collar, with a silver clasp on it, around her neck. He squinted to look closer as she approached, just to make sure it was really what he thought it was. When the look on her face remained serious, his smile disappeared. It was a bold move on her part.

He noticed something in one of her hands, but before he could focus on it, she reached down with her other hand, grabbed one of his, and said, "Come," just like his ex-wife did when he visited her at his old house. After he stood up, she led him out of the living room, and they walked in silence to the door of her bedroom. He could hear the sound of her heels against the wood floor as they moved through the room. When they reached the door, she stopped, let go of his hand, and offered to him what was in her other hand. It was a leash. He looked at it, and then back at her. When he didn't do anything, she attached it to the collar on her neck. Then she offered the leash to him and said, "Take it."

He looked at the leash, and then his eyes instinctively traversed the rest of her body, noticing again her powerful thighs in the tight-fitting skirt. Even with his strong willpower he became aroused, and when he looked back at her face, he saw a woman that was waiting to be devoured. He weighed the offer.

She was taking a huge gamble, and she knew it, but her instincts had always been right, and she convinced herself that he was the type of man who needed to be led, to be guided, while having the feeling of being in control. She continued to hold the leash up for him to take.

Finally he gave in and accepted it, attached to the collar as it was, and led her into the bedroom.

There are those times when it is believed that one is in a position of power and control, when in actuality, just the opposite is true. The one, without being cognizant of it, is the lesser of the two, and is utilized by the other for whatever is necessary and available. For Rand, this was the case that evening. She aimed to exert her control by letting herself be controlled. She believed herself to be exerting influence, and sexual command, over a man who, for all practical purposes, had become immune to such veiled attempts of manipulation a long time ago. She was the lesser of the two.

10

An Old Friend

Things did not sit well with Angstrom for several days after the episode with Rand. It was not so much the after dinner experience itself, significant as it was, but rather, what it represented in the larger perspective of the TTO. She held a lot of power and influence within it, and she was using sex to obtain and maintain that power, bolstering the professional relationships that she had with her client through means other than the evangelization of manufacturing and development processes. How many of them were caught in her web? Rand's grip on the men around that conference table hindered their objectivity. She was maintaining an overly conservative, narrow approach to the crowdsourcing effort, and remaining fixated on the tried and true, the big defense contractors, dismissing outright the technology that Angstrom tried to present.

But however troubling her situation was to him, it did not distract him from what he now deemed his most pressing issue. Even the task of identifying and selecting candidates for the Chassis and Armor category were no longer a priority for him; as far as he was concerned, he was done with that effort by virtue of the PrC submission. His primary focus now was to zero in on and understand what was behind the submission, and to figure out what he was going to do about it. Even though he held the theory of the Chinese scientists, something still made him doubt the veracity of the notion. Therefore, it was a great disappointment to him that he did not receive any meaningful consideration or comments on the PrC technology. That was his biggest frustration from the meetings, even in comparison to what he learned about Rand.

PrC consumed him, like there was some mastermind behind it, lurking in Angstrom's mind, hands around it and manipulating it, massaging it, and not letting it loose; a force pushing him, compelling him to act according to its will. With the full PrC submission delivered, however, the manipulation had been suspended, and its recipient left alone once again, to recoup, to rest, and presumably, to wait; wait for what might happen next.

But Angstrom did not want to sit around and wait; he had to do something, and it was during one of those many restless nights alone in his apartment that an idea came to him. It centered on the phrase at the end of the submission: *For the secret of man's being is not only to live but to have something to live for.*

Perhaps there was more to be learned about it, and he decided to do a search for it on the internet. To his surprise, many hits surfaced for the same, exact phrase. A scan of the search results revealed that it was, in fact, a famous quote. It was from the story *The Grand Inquisitor*, taken from Dostoevsky's *The Brothers Karamazov.*

He wondered why Chinese scientists would quote something from Dostoevsky. Did they believe that their own Chinese literature was too obscure to persons of the Western world, and so they had to borrow something from someplace else? It was a paradox to him, and at times it caused him to question his Chinese theory after the discovery.

He felt strongly that he had in his possession a revelation in military technology. That was why it was so important to have a discussion about it at the Phase-Two kickoff meeting; he was making assumptions about the PrC technology, just like he was about its Chinese origin, and testing that assumption, and verifying its veracity, was imperative. He wanted to hear the reactions of the others and see if they were as impressed with it as he was. He assumed that he had researched it sufficiently so that his assessment of it was an educated one, but just like he had some doubt about his Chinese theory, he had at least a little skepticism about his own ability to settle on the true merit of the technology itself.

Angstrom was not prone to arrive at rash conclusions, and because he was not doing anything but waiting anyways, he eventually formulated a rough plan as to what he could do in the meantime to confirm his feeling as to the legitimacy of the submission, and assuming he got that confirmation, what action he could further take on his own, unilaterally, regarding the whole matter. The first step was to get validation of the technology. If the submission turned out to be nothing, or was a farce—which he doubted would be the case—then that would be the end of it.

He considered many options on how he could get the validation, and he cast a very wide net with respect to the possibilities. He ruminated about past professional acquaintances, other experts in the field; various avenues of thought were pursued, until finally he came up with an idea.

Special arrangements were made for a package to be delivered. He didn't want to leave it up to a normal commercial courier, and fortunately he still had access to the necessary means through his old division to ensure safe delivery of it, along with a short set of instructions on what the recipient was to do with the package (and the time frame involved). A check through an internal data base—he was pleased to find that he still had an account on it—confirmed that the intended recipient of the package had an appropriate security clearance.

A week after the confirmed delivery, on the night before he was to meet with the recipient, he lay awake in bed reflecting upon the conversation that he would be having the next day, formulating his thoughts and line of inquiry. And he thought about the prospect of seeing an old friend that he had not seen in a long time. Before he turned in for the night, he went to his PC and clicked on an image—the second image of Candy Mav. He laughed at her name, but her image continued to linger in his thoughts. Ever since he first saw that second image of her, it stayed with him. He picked up the CD from his desk and looked at her, and then turned back to the screen for another look at her face. Beyond the whore factor represented by her pose, and her heavily painted face with the animalistic expression on it, he detected some semblance of beauty in her. He continued to be obsessed with her, which surprised him when he thought about it, because he was like a schoolboy lusting after the first nude image he ever saw, falling in *love* with it. After a moment longer, he forced the thought out of his head, shaking his head vigorously to facilitate the effort, and did his best to allow himself to relax and fall asleep.

The one hour trip North to Baltimore was a minor inconvenience for him; it was the least he could do in order to seek the help of an old college friend, and for something that was so important. Fred Racal was an esteemed Professor of Electrical Engineering at Johns Hopkins University and was known for his overall expertise in the area of defense technology.

Angstrom poked his head into the office and saw Racal sitting at his desk, studying something fixedly. After a short wrap on the opened door, Racal looked up, smiled, and rose to meet his old friend.

"I can't believe it. After all these years, it's really you—John Angstrom in the flesh. Hello, John," Racal said as he shook Angstrom's hand. Then they laughed and hugged.

"Fred, it's great to see you again. Thanks for making the time to help out on such short notice."

"Are you kidding? Anything that gives me a break from

grading mid-term exams is a welcome intrusion. Come on in and have a seat."

Racal was a distinguished looking man, dressed much more formally than the rest of his fellow professors at the university; it was a carry-over from his time working at a corporation before he had stepped back into the academic world. It was also reflective of his overall character; he was a reserved, serious-minded person, dedicated and passionate about his work. When he was around Angstrom, however, he allowed himself to relax and be much less formal about things. He was never sure why; that was just how it had always been. With Angstrom it was the same way; he was a different person whenever the two were together, able to reveal thoughts and feelings that he would never share with anyone else.

They both sat down and observed each other. Racal smiled in the sheer happiness of the moment, seeing his old friend for the first time in so long, but he could see the wear and tear on Angstrom's face, and in his eyes. He was still the sporting, handsome man that Racal remembered, but he could tell that Angstrom had led a different life as compared to the cloistered life of his own. After another moment he said, "Well, how the hell are you, John? It's been a *long* time."

"Yes, it has. I'm holding my own. Got divorced a couple of years ago, and I'm in a new position within the government that was not on my roadmap, but I'm rolling with the punches."

It was the same Angstrom that Racal remembered, straight shooter, and never beating around the bush.

"Well, you were always a tough one," Racal said, "so I'm sure you'll make it work, whatever's thrown at you." It felt odd to utter such a statement to his old friend; it sounded too cold and distant. It was also strange to him to feel like he needed to give words of encouragement to the man he always felt was on top of the world and in control of his own destiny.

They were great friends back when they were together at Carnegie Mellon—roommates, inseparable, and often taking the same classes together. They were a force to be reckoned with academically; they worked on assignments and prepared for exams together, constantly pushing each other. The academic bar was raised in every class that they took together, and their professors loved the challenge of having them in their classes. Their lives diverged when one chose the corporate life, and then academia, while the other was targeted, hand-picked, and groomed by a highly selective section of the Government, for something very different.

As much as Racal wanted to reminisce about old times, Angstrom was eager to get down to business, and Racal could sense it. He still knew his friend well after all those years.

"Well, I'm looking forward to our dinner tonight," Racal said. "I've got a great seafood place picked out for us, and we've got a lot of catching up to do, but I could hardly wait for your arrival in order to talk about your package."

Angstrom appreciated Racal's overture to dive right into the essence of his visit. "Were you familiar with the TTO's crowdsourcing effort before you received it?"

"Yes. Yes I was—from when it was first announced back in '09. I was intrigued with the concept, and was always curious as to what might result of it. If this package is any indication," Racal said as he pointed to the copy of the PrC submission on top of his desk, "then the TTO is sitting on a veritable gold mine of new technology."

"I wouldn't assume that the PrC submission is representative of the rest of the Chassis and Armor submissions we got; it's a special case."

Racal looked at Angstrom intently, and nodded slowly. "Hmm. I saw the reference to PrC at the end of it; so that's how you're referring to it? P-r-C?"

"Yes. But let's talk about that aspect later—it's certainly interesting in and of itself. What about the technical content of the submission? I'm sorry for diving right into it, but you know me, always cutting to the chase. What's your initial reaction?"

"Yes, well, it makes for some fascinating reading, that's for sure," Racal said as he put his feet up and crossed them on the corner of this desk. "I don't mean to be coy about this, but tell me...because I have to ask: How does the TTO itself feel about it?"

"It's funny you ask, and maybe you could guess the answer by virtue of the fact that I'm here in the first place." There was a slight pause before Angstrom continued. "I floated some of the most interesting concepts to a few of the others at the TTO, and let's just say that I met with some resistance."

"I see. That's surprising. Did you actually show them some of the theoretical calculations and analysis that were included in support of the concepts?"

"No, not really. Not after receiving those initial reactions."

"Then I guess I'm *not* surprised. You see, there are a few things that, in my mind, make this submission incredible. First, it seamlessly integrates a tremendous amount of cutting edge technology, from very disparate technical fields no less, into one hell of a marine amphibious vehicle design. It truly represents the next frontier in twenty-first century amphibious transport. Second, much of that technology, as you're probably aware, has been discussed, if not implemented, to some extent already within various types of military technology. But what this paper does..." Racal said as he lifted the bound submission from his desk and into the air, "is take it all to the next level. For example, the limitations that were previously thought to exist in semiconductor technology, which prevented effective sensor insertion into the armor, are seemingly resolved in this paper. The submitter has introduced groundbreaking work in energy bandgap manipulation, and highly insulated substrates, to produce ruggedized, radiation insensitive semiconductor technology with electron mobilities that are unheard of. That's just one example. The new material science proposed for the composite sandwiched between the multi-layered tiles for the reactive armor...it's brilliant. And third, the submission contains some concepts that are totally new—absolutely revolutionary. I've never heard of them before—never seen anything like it. An example that fits into that category is the proposed millimeter wave vehicle encasement to achieve radio stealth. Essentially, it's an electronic shield generated around the vehicle made up of a high frequency, small wavelength electromagnetic field that renders EM intercept and detection virtually impossible. It's absolutely incredible. I've been thinking about it ever since I read it; the future applications for it are enormous...widespread in a lot of different areas other than just amphibious-land vehicles. And the submission describes new breakthroughs in semiconductor technology to efficiently produce the high powered millimeter waves—breakthroughs worthy of extremely valuable pioneering patents. Whole new companies can be formed around the millimeter wave technology alone."

They proceeded to talk for several hours about the technical aspects of the submission. Angstrom asked many questions, took thorough notes, and sometimes Racal got up and drew diagrams on a whiteboard, along with accompanying mathematical equations, to explain some of the complex, novel ideas involved. Racal had copies of certain journal articles made in advance of their meeting, and he handed each of them to Angstrom at various points during their discussion as background material so that Angstrom could read them later and better appreciate the concepts involved.

Finally, Angstrom leaned back in his chair and massaged his face with both of his hands as he processed everything that he had heard. "So my initial reaction to the document was spot on. This submission," Angstrom said as he pointed to the document which Racal had tossed back onto his desk, "represents legitimate technological breakthroughs."

"Oh, it most certainly does." Racal was standing at the whiteboard, and he looked down to put the cap on the dry-erase marker that he was holding. "I have to admit, I've been dying to learn of the identity of the entity that made the submission. There's a very long conversation that I want to have with them."

"You said *them*. So you believe it's the work of more than one person."

"It has to be." Racal paused and watched Angstrom for a moment. "Surely this isn't the work of just one person, is it?"

"Well, how about we find a quiet corner in that seafood restaurant and talk about it?"

It was a short drive south to the Fells Point area, and it still being daylight, Angstrom was able to appreciate the charm of the inner harbor. They began their walk toward the restaurant on a sidewalk that ran along a cobblestone street.

"I always thought this was a nice area," Angstrom commented as he took in the sights.

"Yeah, me too. There's been a lot of rejuvenation of the area. There's something about it that I find… I don't know…uplifting."

Angstrom continued to look around in appreciation of the area. There were many old brick buildings, small shops, and pubs lining the streets. It reminded him of an old colonial town. He could see, off in the distance, the tops of large ships that were docked at the harbor beyond some of the buildings.

The restaurant was smaller than he was expecting. It was very quaint, but Angstrom figured that they were going to have to be careful and not talk too loudly, or the few other patrons in the restaurant would certainly overhear what they said.

"So, PrC," Racal said as he took a sip of his Chardonnay. "Quite mysterious, coupled with that quote, no less."

Before he drove up north, Angstrom had not really planned on getting into all of the details about how he received the submission; he was just going to focus on a technical discussion in order to validate the technology. After talking with Racal all afternoon, however, he decided that discussing its potential origin might be useful. Racal would view things through a different lens than he did, and perhaps he would see something that Angstrom had overlooked. He was glad that he had left the valediction in the copy of the document that he sent to Racal.

Angstrom took a spoonful of lobster-bisque, and then he said, "So, you think it's a quote then?"

"Of course."

Angstrom waited for Racal to say more. But Racal didn't continue, and instead he just took his own spoonful of bisque—he was toying with his old friend, purposefully making him wait.

PAUL J. BARTUSIAK

When Angstrom saw that he was not going to offer anything further—the same Racal, after all those years—Angstrom said, "I'm still trying to figure out why a team of Chinese scientists would quote Dostoevsky."

"*Chinese scientists?*" Racal said with a slight note of incredulity as he looked up suddenly from his bisque. "Why would you think that?"

Racal had formed his own theory behind the reference to PrC, and he was surprised that someone could have surmised something so different. Angstrom was caught off guard by Racal's sudden, incredulous tone. "P-r-C," Angstrom said. "People's Republic of China." He was almost timid in saying that, if Angstrom could ever be thought of as being timid.

"Hmm," was all that Racal could bring himself to say at first. He shook his head with a note of humor when he realized how someone could have interpreted the initials in that way and arrived at that conclusion. Angstrom was beside himself with anticipation when he saw by Racal's reaction that he had formed a different theory.

"Well, damn it, what's *your* theory?" he finally said in a more flustered tone while leaning forward to make sure what they said could not be overheard. A few other patrons in the restaurant noticed how Angstrom and Racal were leaning forward over the table while they talked, but they didn't think too much of it, because they were engrossed in their own conversations.

"As I think about it now, I suppose it's a bit of luck that this whole thing landed in my lap; a perfect confluence of events, if you will," Racal said.

"What do you mean? Damn it, come on Fred…you're driving me nuts. What the hell's on your mind?"

Racal got a rise out of getting Angstrom agitated—just like old times. "That passage at the end of the submission," he finally said, "It's from Dostoevsky's *Grand Inquisitor*. It's Russian."

"Yes, that much I had gathered."

"And you still got to Chinese scientists…hmm. Do you know much about the story?" Racal asked.

130

"Some. A rough outline based upon what I was able to read online."

"You know, you should really open a book sometime, John. You can't get everything from the internet."

"Fred, I'm warning you."

"Alright, alright. *The Grand Inquisitor*—it's an imaginary tale told by one of the main characters, Ivan, in the novel *The Brothers Karamazov*, and it tells of a second coming of Christ during the height of the Spanish Inquisition." Racal took a spoonful of his bisque and then continued, "Christ is taken prisoner, and the Grand Inquisitor visits Him in prison, lambasting Christ for having originally rejected the three temptations of Satan. Basically, the character of Ivan in the book was grappling with all of the great suffering and problems of humanity in the world."

"Yes, I read about that. So how does that fit in with your own theory about the submission?"

Racal put his soup spoon down and spoke with both of his hands raised over the table so as to emphasize what he was about to say.

"I know of a man that was grappling with such profound issues—a man from another country. I only met him once, about ten years ago, and never saw or heard from him again after that. In fact, he's kind of fallen off the face of the earth. He was a scientist, and I met him at a technical symposium; he told me that it was a rare event for him. His government had never allowed him to travel and attend such symposiums, but in what was a brief moment in his country's history, his government had decided that it was time for the country to once again be known for its great, scientific minds. He was allowed to travel and be one of the presenters at the symposium. I was so fascinated with what he had to say, and that his country had allowed him to speak at the event, that I approached him directly afterwards to ask him some follow-up questions. We hit it off and went to dinner that evening."

Just then, the waiter brought their main entrees to the table. They both leaned back from the table to make room for their dishes to be placed in front of them. On Angstrom's plate were two crab cakes made with fresh crab meat from the Bay, while Racal had opted for Sea Bass that had been flown in fresh that day. After the waiter replenished their wine glasses, they each took a bite of their food.

Racal then continued, "We didn't just talk about his presentation, or even technology in general. We talked about a lot of other things. He talked a long time about the state of Russia and what it was like living there. He asked me a lot of questions about the United States...what it was like, the direction of our society, the freedoms we enjoyed. It was a fascinating conversation."

Angstrom by that time had stopped eating and was listening attentively. "You said Russia."

"Yes, Russia. I'll never forget the man, even after just the short time I spent with him. He was profound in his thoughts—science, the arts, and a deep, philosophical view of what his country was, and what it was becoming. He spoke for a long time about his love of the arts, and especially classical music and literature. He was a great admirer of the works of Tolstoy, Turgenev, and Dostoevsky. He even brought a copy of one of Turgenev's novels with him on the trip; he showed it to me. There were hand-scribbled notes all over the inside of it, like he was studying it as he would a textbook." Racal moved his hand in a writing motion, like he was writing the notes in the book himself. "I'd say he was in his late fifties to early sixties at the time. He held a commanding presence; I was somewhat in awe of him." Racal paused in his story, and each of them ate in silence for a while as Racal recollected the man. Angstrom respected the feelings he could tell his friend was experiencing. Finally, Racal continued, "We exchanged contact information, and I wrote to him a couple of times after that, but I never received a response. Looking back on it, even then he was already dejected about the direction that his country was heading. He all but said so."

Racal had become lost in his recollection of the Professor, and he stared into the distance as he thought back to that day. It made him sad to talk about it, now that he saw the submission, and the quote that was there at the end of it. Angstrom was patient and did not prod his friend. Finally, Racal caught himself and came back to the present.

"And here's why I say it was the perfect convergence of events when you brought that submission to me: the scientist that I had met those many years ago...his name was Professor Romanov Czolski...P-r-C."

Angstrom pursed his lips, and his mind began to process what he had just heard. He looked down at his plate of food, lost in thought. Racal watched him, and he suspected that maybe Angstrom was still not convinced.

"John," Racal interrupted. Angstrom broke from his thoughts and looked up at Racal. "When we were having that dinner..." Racal continued, "at one point in the evening the Professor became markedly dejected. His mood changed. He talked more about Russia—its many problems as he perceived them. At the lowest point of his discourse, he recited that quote from Dostoevsky. He didn't reference it—he just said it, like it was his own notion.

11

Offering the Reins

Legitimacy of the PrC submission was confirmed, and Angstrom was ready to press further. The next part of his plan would be more challenging. Additional caution would be necessary — the utmost caution.

He used a secure line from within the TTO to call into the old number, and an automated response answered. The voice sounded more feminine and human than it used to — the software must have been upgraded, he thought — but it was the same system nonetheless. It was still active, and they had left it open for him, just like they said they would. It would always be there for him, he was told, for the rest of his life, available whenever he needed it. That was the deal for people that served in the kind of role that he had, and he needed to take advantage of that fact just then.

After uttering a select set of phrases into the telephone receiver for the purpose of voice synchronization, he stated the fragmentary words indicating his coded identity. The voice recognition was responding perfectly. Even though the system was designed to automatically identify his approximate location from the packet or circuit-switched data of the telephone network, as the case may be, he stated another code-phrase in order to identify his general location. When registration was confirmed, he verbalized some additional code words, which reflected the specific action for which he was seeking. He was impressed that he still remembered the necessary codes, considering everything he had been through, and the fact that it had been over two years since he last utilized the system.

In response, the system consulted databases, parsed through data, verified calendar dates, and then the soft, female voice issued codes of its own, and Angstrom jotted them down. When he later decoded the instructions, he destroyed the paper onto which they had been written. The meeting was arranged.

* * *

Sunday was selected for the surreptitious meeting. The philosophy behind his organization's standard protocol immediately came to Angstrom's recollection when the day was designated; Sunday was always the best day—no excuses necessary for not being in the office, and it was the day of the week when there was the greatest possibility that those who would be interested in intercepting such a meeting would be attending to their own personal matters. Even spies liked a day of rest.

It was a cool, early spring day, and Angstrom wore a new, beige-colored coat that he had purchased just for the occasion. The trees at Constitution Gardens were still bare, and the wind blew frequently, providing an icy chill. He could see the Washington Monument across the lake. The bright, orange-lit sky made his sunglasses and Washington Senators baseball cap look natural for that time of the year. He looked like an average person, but he didn't move through the park like one; he took a circuitous path to the destination, and then he lingered at a distance in order to exhaustively survey the area before finally walking to an empty park bench and taking a seat.

He knew that the person he was going to meet would not be there yet. Standard procedure was to designate one person to arrive earlier than the other, initially canvas the area, and if anything suspicious was noted, leave, with that alone being the signal that cover was potentially compromised, or at least, the occasion for the rendezvous was not deemed suitable.

It was a remote area of the park, a bench tucked into one of the corners, surrounded on two sides by large shrubbery. He waited for quite some time for the arrival of the other person—longer than was normally allotted—and he thought it peculiar that he had to wait for so long. He looked back over his shoulders and could see no one, save for a couple of people far off in the distance walking their dogs, and he had to lean to the side and almost stand up completely in order to look around the bushes to see them. There was nothing suspicious about them, and he still thought that the area was safe. The female, robotic voice did not indicate how long he would have to wait, because it was left to the other person—the one with whom the meeting was requested—to make that determination.

As he waited, his mind drifted from one thought to another. First it was PrC, then Fred Racal and memories of their time together back in their college days. Those thoughts transitioned to memories of his marriage, and its gradual deterioration over time. His mind worked its wonders by traversing the path from one suppressed memory to another, with each mental jump progressively taking him to a more remote, less connected idea as compared to the original, germinating thought, until finally, the brain recalled something that seemed only vaguely related, if at all, to the initial idea, barely linked as it was by the most minute thread of relation, such that, if Angstrom had the opportunity to consciously compare the very first thought with this last one, he would not have the slightest idea of their correlation. His subconscious reined, and took him to the memory of that fateful day, the one that he had constantly struggled to keep suppressed over the last two years; the one that had caused him to suffer the psychological trauma that he did, thereby necessitating endless counseling with the specialists and experts, and that had ultimately ended his career; the one, terrible event that required and resulted in the separation from what he was.

He had crossed the border from Turkey into Syria, and for one long, hard week, he hid in a refugee camp in the border city of Atma. It was a veritable tent city; the conditions were horrible, with over twenty thousand refugees having nowhere to go, little food and water, no electricity, and about fifty portable toilets to serve the whole population. He was on a one man operation, alone and in deep cover in a remote, semi-hostile area of Syria, his only help being the aid of a local collaborator from a modest, rebel faction. Intelligence reports indicated that a man named Tawzid al-Kallbhair, a leader of an extremist group with purported ties to a terrorist faction, was lurking in the area, looking for desperate souls willing to become new recruits in his group. Tawzid was deemed very dangerous, wanted by the United States Government for his terrorist ties, and his alleged perpetration of horrid acts of terrorism. Angstrom was lying in wait for him.

When Tawzid was nowhere to be found after an entire week of waiting, Angstrom's source indicated that Tawzid had moved into the city of Aleppo, so Angstrom likewise moved into the area. After replenishing his supplies at a military base nestled deep in a fir forest, he again went into deep cover, holed up in a little, one-room hut, with no electricity or plumbing, and he had to lay in wait. He stayed in that small abode for close to two weeks, waiting for Tawzid. The situation was precarious—extremist rebels were everywhere, so it was not possible for him to go outside very often, especially during the day. The overall conditions for him were awful.

When he finally received word that Tawzid was spotted in the city, Angstrom, somewhat weak because of the depletion of his rations, as well as from eating small portions of the bacteria-ridden food that was secretly delivered to him, was still able to act with precision. Based upon the information he was given, he went to the second floor of a vacant stone building in a secluded area near his hut. Tawzid was reported to have been making his way down the narrow dirt alley that passed in front of the building. Looking down onto the alley from where he stood, he killed the three men that were guarding Tawzid with three silenced shots directly into each of their heads. Before Tawzid knew what was happening, Angstrom jumped down and landed on top of him, knocking him out with the blow. He quickly pulled the man inside of his hut, seated him on an old, wooden chair, and bound him. Then he went back and dragged the dead guards into the room so that they would not be discovered. There they remained, just the two of them, one the captor, and the other, the prisoner, with the dead guards thrown in for good measure. He initiated the specialized radio beacon, his only method of electronic communication, to indicate that he was ready for extraction. And then he had to wait.

It seemed like forever to him as he waited for his extraction, the dead guards piled in one of the corners on the floor like rag dolls, with flies collecting around their bodies, and his prisoner bound and gagged, strapped to a chair, watching Angstrom with piercing, evil eyes. Sometimes he blindfolded Tawzid so he wouldn't have to look at him. He thought the extraction team would have been closer so that the wait wouldn't have been so long.

After the first full day of his collaborator not bringing him any food or water, he suspected that something might be wrong, and he became anxious. He certainly could not risk leaving the place, because it was too dangerous. After another day, he began stretching what little food rations he still had, suspecting that he might be there for longer than planned. He could not understand why his rescue had not arrived; the helicopters were only supposed to be about ten miles north, at the Jarrah airfield just outside of Aleppo.

The place began to smell from the decaying dead bodies (the one window of the hut had to remain shut for security purposes, so there was no ventilation). For Tawzid it was worse; Angstrom did not want to risk untying him, so the man had to shit and piss on himself as he sat bound to the chair. The excrement and urine collected on the floor below him. It was horrible.

After two more days, Angstrom became dizzy and feint from the shortage of food, the contaminated water, and the fetid smell. And then he heard a knock at the door—the knock he had expected to hear a long time ago: it was that of the local collaborator who had helped him hide in that room for so long in the first place. He was supposed to have brought Angstrom food and water daily, with the special knock serving as the signal of his arrival. When Angstrom heard it, his eyes looked over at his captive, who himself was woken from his semi-conscious state. Their eyes met, both exhausted, but both suspecting that something was not quite right. Angstrom's ability to think had deteriorated; he was physically and mentally weak. Otherwise he would not have gone to the door like he did, after not having heard from his collaborator for so long. He kneeled down at the door, twisted the doorknob to slowly open the door like he was supposed to, and poked his head out of the crack to look down for his tray of food.

There was nothing there. He looked harder at the parched dirt on the ground, thinking that if he focused better, he might see something, but there was nothing. Then he turned his head to look up, and just at that moment a metal pipe came down and hit him on the back of the neck, knocking him to the ground instantly. The man wielding the pipe dropped it to the ground, un-slung his light machine gun from his shoulder, and hoisted it ready as he pointed it into the door and entered through it, prepared for anyone else that might be inside. The man's accomplice, still standing outside with his own machine gun slung over his shoulder, grabbed Angstrom with both hands and dragged him on the dirt to further away from the door; Angstrom remained there motionless, lying on the ground. He was only semi-conscious, but could hear the two men mumbling to each other inside, and then he heard slapping sounds. They were trying to revive Tawzid and cause him to become more alert in preparation for their rescue.

Angstrom fought to stay awake, still dizzy from the blow. Slowly, he reached into the jacket he was wearing in anticipation of the extraction, took out the M84 that he had in it, and then scooted himself on the dirt, back toward the door. Before the two rebels could see that he was there, he triggered the magnesium-based charge and rolled the canister into the room, and when the two men heard it rolling on the dirt, they both instinctively turned to look at what it was, which was lucky for Angstrom, because their looking at it caused its effects to be magnified. The flashbang grenade ignited and instantly emitted a loud, piercing sound, and a blinding light. Angstrom was not thinking clearly and did not even completely turn away from the blast himself to avoid the flash.

The explosion caused immediate blindness, deafness, ringing in the ears, and general disorientation for all four of the men. The two rebels became confused and lost coordination, falling to the ground and burying their heads and ears in their arms like turtles receding into their shells in order to protect themselves. They had no idea who had thrown the flashbang, and were afraid that bullets would come flying at them at any moment.

Because Angstrom anticipated the blast, he was the first to partially recover; he scooted on the ground to assail one of the rebels in an effort to take the machine gun from him. There was a struggle for it, and the man suddenly arched his back, rose, and flexed his body to throw Angstrom backwards and to the ground.

The other rebel, somehow sensing the struggle, reached for his gun, began screaming, and fired his weapon wildly around the room. As loud as the powerful weapon was, they were all still blind and could barely hear it as it filled the room with bullets. The flesh of Tawzid's face was torn to shreds as bullets ripped into his skin. He was still tied to his chair, and his bloodied, massacred head slumped downward to his chest.

Angstrom, lying low on the ground, leapt and tackled the other man back to the ground and struggled with him for his weapon while the shooting took place. Luckily, the barrage of bullets did not hit him, but he was fighting for his life. The struggle was great, and he feared that the other rebel, the one who had just expended his clip, would soon have his weapon reloaded to fire another round. In addition, all of their senses were slowly returning, and Angstrom recognized it. He drove his elbow into the neck of the man with whom he was wrestling, dealing him a powerful blow, and then another into the side of his face, which caused the man to stop struggling long enough for Angstrom to secure his weapon.

Angstrom's vision was just beginning to return; he could not tell who or what he was looking at, but he saw a blur move in front of him, so he fired at it. His bullets hit the rebel that had just finished spraying the room full of bullets, and he stumbled backward against the wall as Angstrom's gun fired another round that penetrated the man's chest and blasted it open.

The rebel on the floor who Angstrom had struggled with was now on all fours, attempting to get up. While he did so, he grabbed a pistol from the holster strapped across his chest, and prepared to draw it on Angstrom. Angstrom turned to fire on him, and a stream of bullets penetrated the man, starting at his waist, all the way up his body, and into his skull, fracturing it into pieces.

Shell casings were everywhere, and the smell of spent ammunition permeated the air. Everything was silent again; Angstrom's sight was almost completely restored, and he surveyed the room. It was a gruesome scene. Tawzid was long dead; ripped into a pile of flesh, bloodied body strapped to the chair and disfigured beyond recognition. Flies were buzzing around the dead bodies that had been piled up and were decaying in the corner of the room, as well as around the feces and urine on the floor under Tawzid (which was now mixed in a pool of blood). The disfigured bodies of the two rebels that he just fought with and shot were sprawled on the ground, soaked in blood, with more blood splattered on the walls that were riddled with bullet holes. He searched their bodies, checking to see if they had any extra magazines on them, but there were none. He was down to his own G53, which he now found on the floor by his chair where he had left it when he went to answer the door in the first place.

The mission was blown; someone had compromised his cover. At that point, he knew he needed to get out of there, and fast. More rebels would surely come with all of the noise that had been made. He rubbed his head, and then his eyes, to further collect himself and get a better focus on things, and his body began to shake. It was then that he realized that he had been hit—one bullet in his thigh, and another in his right shoulder. He began to feel the pain, and combined with the effects of the flashbang, his starvation and dehydration, and the trauma of the moment, he went into shock; he fell to the floor, bleeding profusely. His sight was becoming unfocused again, and he was losing consciousness. The last sound he remembered hearing, just before he passed out, were the Blackhawks off in the distance; they had received the EM signal from his beacon after all, and they were almost there.

It was his first mission that had ever ended badly, and the only time that he came so close to death. The post-mortem analysis was not kind. His stellar record and reputation were not enough to deflect the criticism.

"He's too old to be in the field...he's left too much to his own devices...insufficient oversight...."

With all of the extensive experience in the field that Angstrom had, including the significant number of important missions, the failed Syrian mission should not have affected him the way that it did. The doctors were not sure what caused his subsequent, mental collapse—betrayal, the near-death experience, the horrific carnage that he was a part of, the cumulative stress of all of his past missions finally taking its toll, his being served with divorce papers (which they eventually learned about), and perhaps just the failed nature of the mission itself—it was all an assault on his psyche. He himself could not get the images of those severely disfigured bodies out of his mind. Tawzid's eyes, looking back at him as he sat strapped to the chair—he could not forget it all. He was declared unfit for further deployment. He thought they meant physically, but later it became clear that it was his mental state that they were most concerned about.

He went through months of physical and psychological rehabilitation, and before he knew it, after all of the therapy was over, there he was, working for Bill Seavers in the TTO. He would no longer being doing what he loved, and the passion in his life was stripped away from him.

A phantom approached, and as it got closer, Angstrom was stunned when he realized who it was. He had never met his handler in a public place before. It was forbidden. They were never allowed to be seen together in public. The man sat down on the bench, separated from Angstrom by a shoulder bag that Angstrom had carried with him as part of his look. They both sat silently for a while and surveyed the area.

A thousand different thoughts ran through Angstrom's mind as he tried to process what was happening. His handler knew it; he was concerned about their meeting, and the emotions it may stir up in Angstrom, but the doctors insisted that by that time it would be all right.

Garrett was tall and slightly heavy set, completely bald, and his cheeks a little puffy with age—he was in his mid-sixties.

After some time had passed in silence, Garrett, looking straight ahead instead of at Angstrom, said in a low voice, "How are you?"

"Are you kidding? What are we doing here?" Angstrom responded in a hushed but nervous tone while looking straight ahead, hands in his coat pockets.

A barely discernible change of expression appeared on Garrett's face, but Angstrom could not see it because he continued to look forward. Garrett said, "You still care about maintaining the public separation between us; that's good."

Angstrom turned to look at him momentarily, but then he turned to look straight ahead again. "Of course I do."

"How's the TTO?"

Angstrom took a deep breath through his nostrils, and steam came out of them has he exhaled into the cool air. He wanted to be angry at the question, but he maintained control and said, "Actually, I'm beginning to adjust."

The amount of words spoken between them was always sparse and direct, and that was especially so in the unique setting that they found themselves. Over the years they had developed an ability to understand each other through carefully selected words, phrases, facial expressions, and body movements. Twenty years together, and an almost perfect match in personalities to begin with, allowed them to communicate almost telepathically.

"Ready to return?"

Another pause by Angstrom as he processed the bomb that had just been dropped. Now he didn't look forward anymore; he turned and looked clear in the other direction; it was out of necessity, in order to remain calm. Finally, he looked forward again and said, "I thought the psychologist's report concluded that that would never be possible."

"True. Not in your previous capacity."

Now Angstrom *was* angry. After all of the debriefing and psychological analysis that he went through, he could not see how the present topic of conversation could even be contemplated. He broke protocol—it was a small fissure, but a break nonetheless: he looked directly at Garrett, in a public setting, and spoke to him. In a hushed, incredulous tone, he said: "What the hell are you talking about?"

He continued to look at the man for a brief moment longer, the longest he felt he could under the circumstances, and then turned to look forward again. It was a signal to Garrett that he was disturbed, greatly disturbed, by their line of dialogue, and Garrett perceived the message immediately.

"Look, I'm in my sixties. I'm done." He leaned forward and picked up a small twig from the ground and slowly turned it in his fingers. "I've only got a handful of clients left anyway...run-of-the-mill. You were the one that occupied all of my time."

"There's still G...isn't there?"

"There is still him; you're correct. But not for me; he was reassigned in anticipation of my...retirement."

There was a long pause after that. Angstrom took one of his hands out of his pocket, curled it up into a fist, and held it up to his mouth as he exhaled warm air into it. He was not prepared for the conversation that he was having. All he had asked for was to make contact with someone from his old organization so that he could hand off the PrC matter to them.

Meeting his old handler, his mentor of so many years, was emotionally difficult for him. It was a shock just to see him, much less under the present circumstances. And then for him to drop a bomb into his lap, just like that; it was difficult for him to process it at first. "I'm done...my retirement...." Those were the words Garrett used, and Angstrom pondered them.

It finally sank in. So that was it, he thought. Garret was offering him the opportunity to be his replacement—the opportunity to become a handler. At first he wanted to laugh—as if the prospect was a sign that he himself was *done*. But then he thought about it, and what he was doing at the TTO, and realized that he was done anyway. After that, the offer did not seem so absurd anymore. If he was going to have a desk job, why not in the line of work that he loved?

When Angstrom did not immediately respond, Garrett knew that he was considering the proposal. Unbeknownst to Angstrom, the doctors had previously concluded that this would be about the right timing to broach such an idea to him; seeing him in person now, and his overall reaction to the proposal, confirmed in Garrett's mind that the doctors were correct.

The idea for his replacement had been with him for quite some time; not too long after Angstrom was pulled from the field, in fact. It was another perfect coalescence of events: Angstrom no longer being able to function in the field, and Garrett becoming ready to step down. Perhaps it was the former promulgating the latter; he wasn't sure, but when he considered his age, and how long he had been in service, he knew it was time.

When there was still nothing further said by Angstrom after

quite some time, Garrett said, "You'll have to train your own protégé, just like I did you."

Still no response from Angstrom, who sat in silence and watched a squirrel wrestle with an acorn some thirty feet away. The concept of training a new protégé was another new twist. It meant that Garrett was serious when he had said that there was no one left under him. No one that was of merit.

"What do you think of him so far?" Garrett asked.

"Huh? What did you say?" Angstrom said as he snapped out of his trance and looked at his mentor for a moment.

Waiting until Angstrom turned away from him, Garrett repeated, "I said, 'What do you think of him so far?'"

"Who?"

"You're potential new protégé; your first candidate."

A puzzled look appeared on Angstrom's face, and he continued to look off into the distance, into a clump of bushes that the squirrel had run into. He concentrated on what Garrett had just said and pieced the fragments of information together. So there was a plan for him after all, he thought. They had not abandoned him.

"You mean...." He stopped from what he was about to say and instead said, "Did you have something to do with his getting his new position?"

A slight nod by Garrett in the affirmative.

"Does he know he's a candidate?"

This time a negative indication.

By then, they had been sitting together long enough such that protocol required another survey of the area; they had to make sure nothing was out of the ordinary. They had both been subject to the rules and protocols for so long, it was automatic for them, and they simultaneously, but ever so casually, performed the task.

Then Angstrom pretended to fumble in the bag resting next to him in order to demonstrate some normal activity, and as he looked down at it he said, "He'll never go for it. He's interested in climbing the ranks of the paper pushers. He told me as much."

"That's his parents talking. We have reason to believe

otherwise."

"Does anyone in my Office know about all of this?"

Silence. No movement by Garrett.

"I'm onto something with what I'm doing over there," Angstrom said.

This time it was Garrett that turned to look at Angstrom for a moment.

In the briefest terms possible, and as cryptically as possible, Angstrom disclosed what had been going on, including his verification of the PrC material with someone on the outside. He spoke in fragments, but the words chosen, and the arrangement in which they were used, were sufficient to convey what he intended.

Garrett expressed knowledge of the situation.

He did not say how he knew about it, but there was no need. Given what Garrett had proposed to him during their meeting that day, Angstrom's communications and actions were no doubt being monitored closely. Such surveillance would have been assumed anyway, in order to keep an eye on the overall mental recovery of one of their old soldiers. But now, with Garrett's proposal on the table, that would have meant that certainty about Angstrom's mental state was even more important to them.

"Did you know it was coming when I was placed where I am?" Angstrom asked. He was referring to the PrC submission.

No response.

Angstrom always wondered how much Garrett knew about certain aspects of his missions, and how much of what occurred was planned, or at least anticipated, ahead of actual events. If he accepted the offer and became a handler himself, he would finally learn the answer.

He put aside Garrett's proposal for a moment and turned his attention back to the PrC situation, which was the reason he had requested a rendezvous in the first place. He needed direction on what to do, if anything, or to know whether they wanted to take over responsibility for the whole situation themselves. It had developed too far for him to move forward on his own, and it was certainly beyond the scope of his responsibilities in his present capacity.

"Play it out, but not on your own; go through 'G.' No one else, except for the one, if you decide to involve him."

Angstrom's mind immediately deciphered: "Play it out" meant handle it, and if it develops into something, fine, but don't let it blow up into a big deal; under no circumstances could it become fodder for the press. "Go through G" meant that Angstrom could not handle field action himself; he was to use someone else in the organization, and that someone was the person referred to as G. "No one else... except for the one" meant that Angstrom was to tell no one, not even Seavers, what was transpiring, except, if Angstrom chose to, if he thought the situation was right, he could involve his new potential protégé, Keplar.

Angstrom considered whether the whole matter was serving as a test for him, or for Keplar, or maybe for both of them. A trial by fire, of sorts. Angstrom was to effectuate Professor Romanov Czolski's defection, and work through Gordon to do so, serving as his *handler* on a temporary basis.

"Six months," Garrett said as he prepared to get up from the bench. "Contact via procedure to let me know your decision. We'll proceed from there. If it's no, it was nice knowing you."

Angstrom took one last look at him—for as long as possible given protocol constraints. It might be the last time he ever saw him.

Garrett stood up, stretched by slightly arching his back to the side, and walked away without looking back. Angstrom watched him as he went down the path, until finally the man blended into the trees and shrubs, and was gone.

He stayed there, sitting on the bench alone, for quite some time after Garrett left, speculating on many things: the proposed succession plan, the temporal relationship between his own mental collapse and when the idea for the succession may have germinated, whether the meeting with Garrett was primarily to address the PrC issue or the succession, and even as to how

Gordon was doing out in the field (and within the organization). Just at the time when the feeling of some semblance of a new life was being established, life-changing events were raining upon him.

12

Clarity...but Obscured

After two years of drifting, a potential path had materialized for Angstrom, and he was receptive to it. Whether or not he ultimately accepted the offer, the mere fact that it was there further affected his mental state. It was not so much a healing anymore, or reversion back to his old self, but rather a progression, or evolution, to something else; the same, but something more.

In thinking about the Professor and his defection, and what he himself would have to do about it, Angstrom already felt a new weight and challenge associated with the position of a handler, and he began to think that he might relish it. Garrett was a master at it; there would be huge shoes to fill.

So little was known about Garrett personally. That was the way the organization wanted it. If Angstrom was ever apprehended during a mission, there would be little that he could reveal about the man directly above him. The interrogators would likely not believe how little he knew, and would probably end up killing him in trying to get what they wanted out of him.

It was the end of a work week, late Friday afternoon, and he was alone in the data room, experiencing these thoughts. He was concerned about what came next with Professor Czolski, if in fact Racal was correct in his assessment that it *was* the Professor that was responsible for the PrC submission (which Angstrom was sure he was). He was sitting in his chair in deep thought, trying to map things out, as if he were looking several moves ahead in a chess match. So intense was his concentration that he didn't notice when someone entered the room; he didn't hear anything and was startled by a tap on the shoulder.

"I'm sorry, John, I didn't mean to startle you," Rand said with a slight laugh.

When he recovered himself, he looked up at her and said, "I didn't know you were in town."

"I flew in a couple of hours ago. I've got to meet with one of the other teams."

"Over the weekend?" he said with a somewhat suspicious tone, which Rand duly noted.

"No, first thing Monday, but there's something else that I have to take care of while I'm here, so I just flew in early to do it over the weekend. What are you doing here so late on a Friday? Found another nudie magazine?" she said as she became more brazen and put her hand on the back of his neck.

He was not sure how he wanted to respond, or even in what direction he wanted to steer the conversation. He was able to consider her in a new light now, given the path that lay before him with his old organization.

He stood up, without jerking away or causing her to think that he was any different, and said, "No. Actually, I was just thinking about some of those technical features of the MAV that I had brought up at the Phase-Two Meeting." While that was not true, he detected the opening to broach the topic and seized it.

"Oh, that," she said in a somewhat disinterested tone. She sat down in the chair that Angstrom had just risen from, and put her legs up on the table, crossing them at the ankles. The bare skin of her legs was on full display for him, which was what she had intended.

Angstrom took note of her new position immediately, his eyes traversing her legs, starting from her ankles, past her knees, and all the way up to her exposed thighs. He could certainly still appreciate her beauty, but he was not interested in her sexually; his interest before was only for professional reasons, and now even *that* was not the case. He did not, however, want her to know that; at least, not yet.

"Are you still fixated on that?" she said. She meant the technical aspects of the MAV that Angstrom had raised.

He put his hand on her knee and began to caress it, looking into her eyes as he did so. Her legs parted a bit to encourage him, and then she grabbed a model mock-up that was within arm's reach, fondling it with her hands while he caressed her skin. His hand moved higher, just within the inside of her skirt, and squeezed the soft flesh of the inside of her thigh. This continued, in silence, for a time.

"Let's go out and have dinner; we can talk about it," she finally said

"Really? You've got no plans for tonight?

She looked at him earnestly, but also with a mischievous expression on her face. He didn't immediately accept her invitation, choosing instead to make her wait, and keep her off balance, but eventually he relented.

When they were seated at the restaurant, he queried her again as to why she was so resistant to the new ideas that he presented at the meeting.

"Oh come on, John, let's not beat a dead horse. I told you, they're not new at all. It's all been tried before, and you'd just be distracting the team and yourself."

He still did not believe her explanation; she had not even demonstrated the slightest interest in looking at some of the supplementary material that he offered in support of the new MAV concepts.

"Look, John. I've been doing this for a long time now, and I know how those boys think. They're jumping through hoops as it is with the whole crowdsourcing notion, and they don't want to take any more risks than they have to. If they go into unchartered territory with nothing to show for it—except for just a waste of a lot of taxpayer money—they're not going to be around for very long, and they know it. This project is getting a lot of press; any missteps will be magnified tenfold."

"So you don't want to take any risks? That doesn't seem like the Susan Rand that I know," he said, in a playful, teasing tone.

A smile appeared on her face when she considered the double entendre of his statement. "I love taking risk, John," she said as she leaned toward him over the table. "Can't you tell?" Just then she reached with her foot under the table to rub it against his leg. She stopped talking and took a sip of her wine, watching him as she continued to caress his leg. Then she put her glass down and said, "Let me put it another way: I'm averse to failure. Just play along, and when the obvious choice is selected as the winner, be happy with the result; clap with everyone else, and pat yourself on the back. Quite frankly, everybody's job will be easier, yours and mine included, if we don't have to dig ourselves out of some hole. Why do extra work if we don't have to? When the project is successful, everyone gets rewarded. You get a promotion, I get another project, and everyone is happy. Some day you'll leave your government position, wait out the "cooling off" period, and then get all kinds of job offers from the big defense companies. Everyone leaves the party happy."

Angstrom chewed his food in silence as he listened to her. She noted how quiet he had become and wondered for a moment whether he disagreed with what she just said. She decided to switch gears and take a different approach.

"Besides, in case you haven't noticed, I move in high circles within the TTO...with people who control careers...people who can either help you or hurt you—TTO, military, ranked officers, you name it." Angstrom understood her meaning, and as soon as she said that, she noticed a subtle change in his countenance, and she became concerned that she may have gone too far, misplaying her hand as it were. Angstrom put her off of her game—she was beginning to realize that.

He still did not say anything. He had stopped eating and sat there watching her with both hands resting on the table, holding his fork and knife steady.

Looking at him, she began to think that he looked vulnerable, like maybe she *had* struck a nerve. Reaching across the table, she grabbed his hand and said, "Just don't fuck this up, alright?" She squeezed his hand, trying to convey some manner of control over him. Her message conveyed, she changed the subject and said, "Let's go to your place after this; I want to see where you live."

She was all over the map as far as he was concerned, and taking her to his apartment was the last thing on his mind. He wanted to run, and run fast. He figured that he had gotten about as much information as he was going to get out of her, and it was nothing too earth-shattering. He did not, however, want to upset her or let her know his true feelings, and he was curious to know how much further she might try to go in order to get him under her spell. After all, she had gone quite far already. He decided to play along for a little longer.

"No, not my place. The maid was sick and didn't get a chance to service it. Let's go to your place."

"My place is…boring. I don't care about a little mess; let's go to your place."

For some inexplicable reason, it was at that moment that she began to consciously notice that there was something different about him: a change in him since they had last met. He was more confident, and relaxed. She was confused and couldn't read him, vacillating between different notions by the minute.

"Next time—not tonight. It's your place or no place. Sorry."

"Fine," she said, in an overly exaggerated manner, almost with a pouty look on her face. Angstrom was amused, marveling at how such an act might otherwise work so easily on someone else.

When they arrived at her suite, she started to attack him sexually before he even had a chance to take off his coat and sit down. He kissed her in return, almost in self defense more than anything. She grabbed his hand and led him back to her room again. Maybe there would be no leash this time, was all that he could think.

She wasted no time. She kicked off her pumps and began to take off her earrings as she looked at herself in the mirror. He sat down on the edge of the bed and realized that it was a mistake for him to have let it go as far as it had. Now he wondered how he was going to get himself out of there. All of a sudden, something hit him on the thigh and startled him. She had tossed something onto the bed.

"Here," she said. "I'm going to go to the bathroom to freshen up...and put on something I think you'll like. You're going to make use of that on me when I get out." Before he could say anything, she turned around and went into the bathroom.

He looked down on the bed to see what had been tossed at him, and he couldn't believe what he saw; it was a horse whip. He looked up at the door she had gone through and was left to wonder at the sexual fantasies she played out, and how they must have ensnared her other prey. He could not even tell if what she was doing sexually was what *she* wanted, or whether it was what she thought *he* wanted, and he laughed at the absurdity of the whole situation. In a way, he felt sorry for her. It was pathetic to him. In fact, as he thought about it further, he realized that it was an odd coincidence: sexually, she was the polar opposite of his ex-wife. Breann liked to control him, whereas Rand liked to *be* controlled—she liked to be told what to do (or so she led him to believe). She had no sexual hold over him as it was, but that recognition of Rand, in comparison to his ex-wife's preferences, further dispelled any chance of his being genuinely attracted to her, and in that instant, he became convinced that he needed to get out of her apartment.

"Susan, I need to go," he shouted to her in the bathroom.

"What?" she said through the door. When she heard that, her heart sank, and she felt foolish. She was not sure whether she had pushed things too far or not, and she poked her head out of the door and said, "Why? What's the matter? Is everything alright?"

He stood up from the bed. "Everything's fine. I just got a text message from one of my daughters on my cell-phone," he said as he held it in the air. "Much as I hate the prospect of having to leave, I need to get a signature to her, ASAP. Children have a wonderful tendency of springing things on you just at the right moment. I'm sorry…" he said hurriedly as he began walking out of the bedroom.

The quicker the better, he thought, and he didn't even wait for her to dress herself or come out of the bathroom.

Before she knew it, he was gone. She couldn't really tell what happened, or whether he was telling the truth or not. She sat on the edge of her bed, right where Angstrom was, and thought about him for a while, trying to figure it out.

She wasn't ready for the evening to be over, charged up as she was. On a whim, she picked up the phone and dialed the number of another new member of the team. He accepted her invitation to come over without hesitation, and she told herself that she would answer the door when he arrived dressed in the leather outfit that she was now wearing. "We'll see who's afraid of taking risks," she said to herself, and she was lustful in her anticipation of what was about to come.

The following Monday, Angstrom went to Seavers' office. He informed him that things were on track with regard to the submissions, and Keplar was carrying the ball on much of it. He also indicated that, without being able to say why, he would not be around as much, and there could be stretches of time when he would not be around at all.

Seavers nodded in acknowledgement, saying he had already received such an indication from high up the chain of command.

* * *

The Professor knew that the noose would be tightening around him after no one showed up at the café. He tried to explain away the discrepancy to Dmitri, but it was doubtful that the cancer blinded him anymore. Not after such heightened expectations in anticipation of the first meeting, only to be stood up when neither the Professor nor his guest of honor made an appearance.

Beak-nose, as he thought of him, would not take it kindly either, having made such a public appearance only to realize that he himself may have been the one being scrutinized. In the world Beak-nose dealt with, there were all kinds of excuses proffered for failed meetings—flat tire, missed train, oversleeping—but they were never believed; there was no such thing as a coincidence. He would berate Dmitri; he would put him on notice that the plan, the Professor's plan, was not proceeding as it was supposed to; the Professor was holding one over him and not sharing all of the details.

The Professor knew that Dmitri would not handle Beak-nose's pressure very well. He went into deep thought about it. Dmitri, he wondered to himself, what could have brought you to this? What hole in your life are you trying to fill? Can't you see what you're bringing upon yourself? They were questions the Professor pondered over and over again.

He had always assured Dmitri that he would look out for him, and that *something* would be there for him, come what may. That Dmitri would betray him as he was, then, was a dagger to the Professor's heart.

Dmitri himself was not quite sure why he was doing what he was. He did feel some bitterness about the Professor's life and career. The Professor, he felt, rode the proverbial back of the Russian military complex, rising in his career through extensive, government-funded research on military technology. The Professor's stature grew along with the growth in military research expenditures. Now that the military spending was drying up, there was none left for Dmitri and his generation. The Russian Government was making it a priority to cut military technology research and development. "Use what we have," they would say, "and for what we don't have, buy from other countries that are spending their precious dollars on new weaponry." So where was the opportunity for Dmitri? Where was the overabundant and free-flowing government spending that Dmitri could tap into and build a career upon? He resented, as he saw it, the opportunity that the Professor had to ride the wave and grow his career, when those similar opportunities were not there for him.

And Dmitri, despite the endless conversations he had with the Professor over the years, felt no affinity for the views on Russian culture and society that the Professor held. He was of a new generation, the short attention span generation, the impatient one, with diminishing appreciation for the past. He embraced modern Russia and turned his back on the staid mannerisms of Russia's past. And if the poor became poorer; if the proletariat were trampled on at the expense of the corrupt—the graft, racketeering, and government theft—that was of no consequence to him. Something in the way that he was raised made him ignore the plight of those less fortunate than himself, as if they alone were responsible for their own misfortune. Thus, Dmitri saw the Professor's public outcry against the Russian government as the furious exclamations of an out-of-touch elitist, and he was deeply concerned that those rants could result in Dmitri's losing what little privileges he had obtained up until that point.

But, even more so than anything else, at the essence of Dmitri's betrayal, and the underlying impetus for his overall negative outlook on life, was a deep-seated resentment of people, and the world in general, which he harbored due to the ugliness that he experienced as a child. The Professor and his wife could not possibly have foreseen how Dmitri's painful, early childhood experiences, stemming from an abusive father and alcoholic mother, would ferment within him, and psychologically scar him for life. Nor how, when that resentment was mixed with and placed in the setting of the corrupt environment of present-day Russia, it would cause Dmitri to form into a being that was cold, calculating, and ruthless—and to cause him to ultimately try to take, through whatever means possible, everything that he desired. The greedy, envious, and destructive nature of his soul went undetected by the Professor and his wife when they took him under their wings. They had no idea of the disturbed spirit that they were bringing into their lives.

The profound anger, and burning desire to rise in social status, was a veritable sickness within Dmitri, in his very soul, which he could not, by himself, eradicate or remediate. It festered within him and rotted his insides—an evil spirit whispering to him in his mind at every juncture of his life. To view it objectively, and in context of the Professor and his now-deceased wife, and all of the nurturing, love, and attention that they provided to him, and the fact that he still turned out the way that he did, begs the fundamental question: Was the evil always there to begin with?

The final tipping point of it all, the factor that finally pushed him over the edge and urged him into the current course of events, was the Professor himself. In him Dmitri eventually saw an opportunity—a path to his aims of success, fame, and fortune—and the path seemed an easy one. For the Professor was too good of a man; his kindness, good intentions, benevolence, all of it, was the opening Dmitri needed in order to reach his distorted, black dreams. And when, in Dmitri's opinion, the Professor had gone too far in his rebellion against the elite of Russian society, Dmitri told himself that the time had come to act; to do something; to leverage the old man in whatever way he could to facilitate his achieving what he himself desired.

And with regard to the Professor's grand plan, Dmitri feared that whatever his situation may have been in his own country, and in his profession, whatever status and success he had obtained thus far, it would all come crashing down if they *did* defect to the United States. They would be gone from Russia, and Dmitri would have to turn his back on everything, whatever that everything encompassed in his own mind; all that he had worked and struggled for would be lost. The Professor—that selfish, out of touch one, as Dmitri began to feel about him—was going to cause him to lose everything and make him start all over again in a new country. A country that he thought would actually despise him and the Professor, and just use them to get what it wanted, and then spit them out.

"Trust me," the Professor had told him. "It's a great country, and our reputations (or rather, the Professor's reputation, if the Professor spoke honestly) will ensure that we are provided for." Dmitri doubted that. Why would the United States set them up and spend its precious dollars on an over the hill professor and his assistant? In Dmitri's mind, the postulate was farfetched at best; more likely they would be setting themselves up for long, arduous interrogations in a dark room with a hot, bright light shining in their faces, and then, nothing—nothing but obscurity.

A chance meeting with Raskalnikov (both were in attendance at one of the Professor's infamous speaking engagements), and the ensuing conversation between them, caused it to become apparent that the two had quite common interests at heart.

So the Professor had lost Dmitri; that he now knew. He wondered at what point in time it was that he *had* lost him. It was no matter, he ultimately concluded, because whenever it was, and for whatever reason (or reasons), it happened. Thus, his world was becoming even smaller and closing in on him. Those close to him, including those he *thought* were close to him, were disappearing from his life. But he would not let it deter him. His singular focus was to succeed in his plan. He had to…for *her*.

When he read the letter from his University that was delivered to him at home, he knew things were going to get bad very quickly. The administration informed him that he would not be scheduled to teach any classes the next semester. He would not even continue to receive a small stipend; he was banned from the campus starting the next semester.

It would be a long, slow descent for him. They, the Russian power brokers, the KGB, the Kremlin, would watch him shrivel up and die, like a bug withering under the heat of the scorching sun, its rays intensified by a magnifying glass.

He knew that soon his internet access would be totally disconnected at his home, and whatever they were letting him have at the University during the current semester would be cut off. That was certain. His communication, and his link to the outside world, would become more tenuous by the day.

That all of his extra planning was prescient was now confirmed in his mind, for he had arranged for other students—students whose talents he could leverage in the event that he could no longer rely upon his protégé. Their work was funded through the aid of his professional colleagues, those other few members making up The Mighty Five. They allowed him to repatriate, to siphon-off from the top, a small portion of their own research funding, and use it for his special *Projects* with those students, his worker-bees.

As for those students, they had no idea as to what real aim their efforts were being directed, because the work was fractured, and pieces of it were distributed to each of them individually. Taken in isolation, each small project, spread across the students of several universities, seemed innocuous enough and could not be linked to any of the other pieces.

As to the glue that would put it all together and make it work, that was through the efforts of the Professor himself. He had rolled up his sleeves to write certain blocks of code. He had to; there was no other way. He had been teaching at such a high level for so many years, however, directing the work of others rather than actually doing any work himself, that at first he wondered whether he would still be able to actually do the work.

But he was still a seasoned professor, after all; they had not taken his mind away from him—not yet, anyway. He used the technique of steganography in ways never done before, and provided patches of code to the other students that had embedded within them new patches of code and functionality. It was sophistication at its acme, patching together the various code modules from the students with his code plugs, and then that code itself having hidden within it, different functionality altogether. It was a new form of steganography with the computer code itself; hidden code embedded within the baseline software. The interfacing of it all, and developing the programming interfaces to source it, was a masterwork. It was beyond cutting edge computer science, breaking new ground in the area of covert, microsourced code development.

And it was through those means that he had instantiated, in some remote servers in the Ukraine (routed through his own complex, steganographic means via Dmitri's servers in Kazakhstan), the code plugs necessary to carry on further communications with his audience in the United States.

As far as Dmitri was concerned, each step that he and the Professor took would occur according to their agreed upon plan. However, each communication with the United States that had the Professor's and Dmitri's instructions steganographically embedded within them would be further coded, unbeknownst to Dmitri, with the Professor's distinct, different instructions further embedded within that. Dmitri would believe that things were going according to plan, all the while the Professor would actually be directing something else to occur according to his own secret plan.

The possibility of Dmitri's betrayal was a contingency in the decision tree that the Professor had fully accounted for. He had to; there was no choice. All the possibilities had to be accounted for.

All that was necessary to activate this alternate plan, to wake up the embedded sleeper code, was for the Professor to make a one-time access to a very obscure website, and then input an equally obscure alphanumeric string. That would cause the first multi-threaded steganographic transmission to be sent to his American listener. The message would be automatically linked to the general IP address of Angstrom's home network and tunneled through the very router in his home to the PC that had been utilized to access the Candy Mav web site.

He rushed back to his home after the scene at the café and woke his tablet from its sleep state, nervous as to what he might find when he tried to access the internet. He hoped that from the point in time when he had left the scene in front of the café up until now that the internet connection to his home had not yet been shut off. He needed a connection to it one last time in order to access the obscure web site and enter the alphanumeric string.

After typing in the address, he waited for the site to come up, and it did! He had made it home in time! Now, with the alphanumeric string entered, all he needed was confirmation of its receipt—a message indicating that transmission of the string had successfully reached its intended destination. That message would be implicit, in the form of a redirection to a certain, predetermined web site. When that redirection came, he could not have been more relieved; he saw the address of the web site appear in his address bar: *http://www.mariinsky.ru/en/*. It was the web site of his favorite theater. Alphanumeric transmission successful!

* * *

Angstrom thought for days about the initial contact that he would make with Gordon, but more so about what they were going to do once that contact was made. He had no plan, and could not make one—not until he knew what the Professor wanted to do next. He checked his home computer every night to see if some communication might have been sent to it, and he reviewed the submission over and over again, poring through it looking for something, some clue that might have been there all along and that he just missed, some indication as to what he should do next, but there was nothing.

Then one evening while he sat alone in his apartment, surfing the web and doing some further research on the new technology disclosed in the PrC submission, his computer stuttered. A window opened up; it was a command window, with a black background, and a prompt waiting for input. He wondered if he had accidentally hit a key, or whether the popup blocker failed to block something. Program instructions began typing themselves at the command prompt; they were commands he had never seen before:

> *Hyper-tunnelRAMsteno.exe...IPconfig.rendermaxiout.exe ...server-ping-verify-retrieve...*

The commands were entering themselves so fast that he could only make out bits and pieces of them, but he could tell that they were invoking programs on his computer; programs, he now realized, that were previously stored on his PC without his knowing about it, unblocked and undetected by his security software.

He knew what was happening. It was the Professor doing his magic. He sat back and left his computer alone as the remote programming did its work, and he marveled at the efficiency and power of it; complete control of his computer was lost. Status bars occasionally popped up and showed the status of certain actions occurring, filling up quickly and then disappearing almost as quickly as they had appeared. Multiple command windows opened up simultaneously, one layered on top of the other, each performing a designated sequence at a feverish pace. Status bars progressively filled and then disappeared, and the process repeated itself.

As suddenly as it had all started, it stopped, and his computer went idle. Then, just when he was about to reach up and tap something with his mouse, the computer became active again. He leaned over to look at his computer case sitting on the floor below his desk and saw a red LED flashing; his hard drive was accessing something. The final result was an information packet instantiated onto his desktop, pieced together by all of the strings of furtive code that the Professor had patched together (Angstrom no longer even had to manually invoke the PrC_STEG_Extraction program; it was all done for him automatically). The packet comprised a set of instructions for Angstrom, along with a brief introduction to the creator of those instructions: Professor Romanov Czolski.

13

Definitive Phase Two

It was the first time in two years that Angstrom had set foot in his old organization's headquarters at Langley. It felt strange to him, walking again in the halls of the place that he had lived in for so many years. People recognized him as he made his way through the area and were stunned to see him after being gone for so long. They stopped what they were doing and watched as he passed through the area. It was like he had come back from the dead.

"Why can't I talk to Gordon directly?" Angstrom asked Tom Franklin, incredulously. Gordon no longer reported to Garrett, and instead had Franklin as his new handler. The organization was already putting the steps in motion in anticipation of Garrett's retirement.

Franklin had been a field operative himself, a former peer of Angstrom and Gordon, having only recently been promoted to the role of a handler. Although Angstrom and Franklin were peers in their former lives, Franklin had never reached the stature or gained the reputation that Angstrom did, and a feeling of inferiority was influencing the way that Franklin interacted with

Angstrom as a result. Franklin wanted to exhibit an aura of authority in his new role.

Angstrom could anticipate already how Gordon probably felt about this new reporting structure and laughed to himself as he thought of Gordon's predicament. Gordon and Angstrom were good friends, about as friendly as two field operatives could be, and they greatly respected each other professionally. Reporting to Franklin would be the last thing Gordon would want.

"He's in deep cover in Europe. We can't pull him out or it would blow it. You know that's how it works. Look, I'm happy to loan him to you—well, maybe not that happy—but I'm willing to loan his services out in light of the transition Garrett is experiencing, but you'll have to communicate with Gordon through me. That's the only way; we've got an intricate network set up for him, and I'm not willing to have it compromised."

Angstrom stood in Franklin's office in silence, processing what he had just heard. Franklin, watching him, thought he was the same as he remembered him, a man of few words, overconfident, and with notions of superiority.

Resolving to himself that the situation was what it was, Angstrom took a deep breath and sat down, placing a small folder on Franklin's desk. He opened it and pulled out an eight-by-ten black-and-white photo printout. "This is the man we're going to get out. His name is Professor Romanov Czolski."

He slid the image across the table, and Franklin picked it up to study it. It was a profile image of the Professor, from the shoulders up.

"Are you sure this isn't just an old picture of Karl Marx or something?" Franklin said with a sarcastic grin. When Angstrom did not respond, Franklin continued, "I mean the hair, you know, does anybody really wear it like that?" When Angstrom still did not say anything, Franklin regretted saying what he had. He had slipped into his old manner of speaking without reserve, and it was unprofessional, and worse, not reflective of the stature of the position. He knew it, and became chagrined with himself for the slip.

"He's an esteemed Professor at a technical university in St. Petersburg. More importantly, he's a military technology genius, and he wants to defect to the United States. He's already provided us with a gold-mine of new technology, and we're eager to have him come over."

"How'd you get this image of him," Franklin asked, holding it up, "or how'd the guy even make contact with you?"

"That's a story for another day. I'm in contact with him, that's for certain, and he's setting plans in motion for his defection."

"What do you mean *he's* setting plans in motion? What does he need us for then? Shouldn't *we* be making the plans?"

Angstrom took a deep breath. He was not used to being patient and having to explain himself like he was. Back when he was an operative, he told Garrett what he was going to do, or what he needed others to do, and that was it; Garrett ran all of the interference and went through the appropriate channels. Now it would be different, and Angstrom realized that. If he was going to be a handler, he was going to have to communicate with other handlers—massage egos, explain things plainly—and it was something that he would have to force himself to get used to, and to develop the patience and tolerance for.

Finally, he said, "Look, this guy's a genius in more ways than one. He's devised ways to communicate with us covertly, and undetected by the KGB, through means that we've never seen before."

Franklin became more upset because he felt like he was being kept out of the loop; he had not seen or heard anything about these new, covert communication techniques. Before Angstrom could continue with what he was saying, Franklin couldn't resist the temptation to interject, "Well, you've been out of the field for too long."

Angstrom ignored the comment.

"I'm getting information in a staged release fashion, and at each step of the way, it's according to his schedule and initiative. So far it's worked. When we finally get him in, we're going to have whole debriefing sessions dedicated to his covert communications techniques alone. It's beyond state of the art."

After a slight pause without Franklin responding, Angstrom added, "I trust what's going on, and I trust him."

Franklin was hot, and he said, "Well, that's fine for you to trust, but when it involves someone on my team, that's a different story. I mean, for all I know this could be a trap to fish out our deep cover over there."

That caught Angstrom off guard. He had never considered whether this could all be an elaborate trap. He wondered for a moment whether he had become soft and careless after having been away from things for so long. But then he thought back to Racal's analysis of the PrC submission, and how groundbreaking he confirmed that it was, and he knew it was not a trap. The technology was too important—too valuable—to expend on something like that.

Angstrom looked at Franklin and again considered the new organizational challenges that he was going to face as a handler. Challenges that Garrett had to deal with for all those years and that Angstrom was oblivious to: other handlers struggling to assert their own power and authority, red tape, egos, and general ineptitude.

Angstrom had no patience for it, and he decided that if it was necessary for Franklin to know more, the Director would have told him more. The fact was, Gordon was assigned to work with Angstrom, Angstrom had been offered the position of handler, and Angstrom was assigned to be the lead for the PrC matter. With that in mind, he looked Franklin squarely in the eyes and said, "Look, is Gordon going to be able to support this or not?"

The direct approach had its intended effect. Angstrom's reputation, and his sudden directness, knocked Franklin back. Far be it from Franklin to doubt the wisdom and authority of the Director.

"Yes, he's your man."

"Good. Now where is he?"

Franklin hesitated, reluctant to divulge the information. He was fighting those last remnants of resistance that urged him to withhold information, but he finally gave in and said, "He's in Helsinki."

"Deep cover in Helsinki?" Franklin did not respond. "How are you communicating with him? Direct or indirect?"

"A little of both. Some of the direct is through rotated web sites with coded content."

"So he's got access to electronics; good. This will be perfect. He's going to need to move into St. Petersburg. Is that possible?"

"Yes. Yes it is. The groundwork is in place."

"Good. Here's a web site. Get him to access it *after* he gets into St. Petersburg. It's got to be afterwards—that's very important." He was concerned that what he had just said did not register with Franklin, but he took a small piece of paper, wrote down some information, and slid it across the desk over to him.

Franklin looked at it, then at Angstrom, and then picked up the sliver of paper and studied it. It was a website with a Ukraine country designation.

"The Ukraine? What's he going to find on the site?"

Angstrom did not say anything at first and instead just looked at Franklin. He was reluctant to divulge too much...even to another handler. "Don't try to access it yourself because that would disrupt the site. It's expecting access from within St. Petersburg. Anywhere else will cause it to shut down. Can Gordon still get online once he gets into St. Petersburg?"

"Wait a minute. He's got access in Helsinki, but St. Petersburg will be more difficult. That would put things at greater risk."

"It's the only way."

"Hmm. Well, we can get him access on a very limited basis, for a short time, and only once, when he's in there. After that, any remaining communication will be very difficult and will need to go through me."

Angstrom was satisfied. He did not agree with the last part of what Franklin said—there was no way he was going to communicate through Franklin—but he could deal with that later. The important thing for now would be to get Gordon into St. Petersburg and get him to access the Ukrainian site.

"By the way," Franklin interjected, "What do you mean *it's* expecting access? What *it* are you talking about?"

"That's complicated. Just get Gordon into St. Petersburg and have him access the site."

Franklin was furious about not knowing everything—not being told the details.

* * *

Amid all of the intrigue occurring on both sides of the world, somewhere in Washington, D.C. there was a different kind of intrigue unfolding; it involved a certain woman and her new prey, and the procurement of control through the appearance of ceding control. The initial seed had been planted with an adjustment of his collar, and then later, the initial stroke on the cheek. Now it had fully germinated.

It was new to him; he was plunging into the depths, and he knew it. He wanted to stop himself, but he could not turn her away, or resist her enough to walk away himself. She was too beautiful, too experienced. She asked to be bound, and he complied. He complied with everything that she asked for, because he was caught. The weakness in him, once detected, was being preyed upon to its fullest, almost as if to spite the one who was really overseeing him, and who had spurned her. Lying on her back with him on top of her, she spotted a small tattoo the size of a thumbnail on one of his shoulders as he maneuvered. She said in surprise, "Hey, what's this?" Her head bobbed up and down slightly as she studied it. Then she kissed it—just before they further submerged into one another.

Keplar asked for something new of her that evening: he asked her to keep wearing her red-rimmed glasses. For some reason he could not explain, it felt kinky to him; almost as kinky as some of the other things that they did. It made him feel like he was doing it with his third grade teacher that he had a crush on as a child, or something to that effect.

For her it was a significant development. She had gotten him to take the next step, the next plunge, by getting him to start making his own requests—to reveal one of his own fetishistic tendencies—and it pleased her. She knew at that point that she had him, completely. Now it was just a matter of her getting pleasure out of it.

* * *

Gordon was stunned after having checked in at the established time via the sat-com link only to learn of the new mission involving Angstrom. The name was referenced as Applecore over the encrypted link, but he knew who that was. Angstrom always went by that name; it was his designation over the link.

They had been through so much together, on so many assignments, and now, Applecore was back. Gordon couldn't wait to return to the States and get the whole story.

His immediate concern, though, was to cross the border from Finland into Russia and access the Ukrainian web site. Why he had to wait to access the site from within St. Petersburg was beyond him—it added significant risk to the mission—but those were his instructions. There was a small, seldom utilized safe house set up within St. Petersburg, maintained at an extremely low profile in order to minimize risk of discovery, and that was his destination.

Getting onto the Allegro bullet train at the station in Helsinki would not be a problem, and the two-hundred kilometer trip would not be bad either. He would spend a good portion of it in the dining car. It was getting off of the Allegro at the Ladozhskaya Station in St. Petersburg, without blowing his cover and alerting the authorities, that would be more difficult. He was going to have to call in a favor from one of their collaborators. He hated having to use one; they were so hard to come by now—each one was vital.

He was not sure how much time he had to get where he was going and do what he needed to do, but he wasn't going to waste any time. He took the five-twelve train on Sunday morning. It was one of the soonest available, and traveling that early on a Sunday would mean the least number of officials working at the Ladozhskaya Station.

After finding his seat, he didn't wait long before heading to the dining car to order breakfast. There was no way that he was going to ask for an English newspaper; he did not want to be noticed or stand out given the sparse number of passengers aboard. When he ordered from the menu, he didn't even bother to speak, and instead simply pointed to the items on the menu that he wanted, all so that traces of his foreign accent could not be detected. As he ate his breakfast, he looked out of the large, panoramic windows at the beautiful pine forests and small Russian villages that passed by.

When the train arrived at Ladozhskaya, he cautiously followed the other passengers toward immigration. It would be difficult, if not impossible, to slip beyond the rope and peel off from the other passengers to bypass the process, because there were not enough people around that early to obscure his attempt, and he would have been easily spotted by the authorities. He was going to have to go through immigration.

He followed the crowd toward the several lines that formed, but instead of getting into one, he went straight toward the aisle that was dedicated to train personnel. No one was lined up there to go through that checkpoint, and when he was about ten feet from it he stopped, stood behind a concrete pillar, and watched the person manning the checkpoint. He checked his watch for the time, and then turned back to observe the person. After a couple of minutes, Gordon saw another worker approach that person, whisper something into his ear, and then switch places with him. That was it; the switch had been made on his behalf. He casually, but quickly, walked to the checkpoint and flashed his fake passport. The person looked at it, then at Gordon, and then nodded his head for Gordon to pass through. Gordon walked past without saying a word, and when he was about thirty feet beyond, he matter-of-factly looked back at the worker, who at that moment was switching places again with the person he had temporarily replaced.

There would be a heavy payment due for the favor, and Gordon would make sure it was paid. Paying off such debts was important in order to increase the likelihood that all such transactions would be successful in the future.

When a taxi cab dropped him off at his requested destination, he stepped out into the cool air. It was late morning, and people were beginning to move about the city, which was good, because it helped him blend into his surroundings. He walked for a while and tried to look as casual as possible, observing the area and endeavoring to note any peculiarities. When it appeared that no one was following him, he headed for the house.

It was a small, old house of wood construction on the outskirts of the city. He purposefully passed the house and continued walking down the sidewalk for a few more houses, and then he turned and cut through two houses to head toward the back of the safe house. A key was lodged in a rotted wood board on the back porch, just like he had been instructed it would be when he was briefed.

The shades were pulled down so that no one could look in from the outside. After entering, he briefly looked around, and except for a few items of furniture in each room, it was essentially empty. There were a few cans of food in the cupboard, running water, and that was about it for consumables.

The most important and immediate necessity was internet access, and even though he had been promised that it would be there, he still had his doubts, not only because it was generally difficult to obtain within Russia, but also because such a modern convenience would certainly seem out of character amongst the old and tattered furnishings sparsely populating the inside of the house. He climbed the unadorned, wooden steps to the second floor, and then he reached for a string dangling from the ceiling to pull down the attic door, along with a set of fold-down steps.

The poor ventilation in that dark, dusty attic made the air feel stale. He pulled on a thin, metal chain to turn on a light attached to the ceiling, which made visible a small table with a laptop computer resting on it. So there was a PC after all, he thought, and it even looked new enough to be useful. He went over to it and turned it on, and it worked. There weren't too many programs loaded onto it. Most of the software applications, save for a word processing application, had been stripped from it, leaving it bare so as to serve mainly as a communication terminal for accessing the internet. He noticed that an Ethernet cable was attached to the back of it and was fed through a hole that had been drilled into the ceiling below, in between two wooden joists. So much for aesthetics, he thought.

After verifying that he truly had internet access, he pulled out the small piece of paper onto which he had written the web site and typed it into the address bar of the browser. A command window opened on the display, and several commands began appearing and entering themselves into it. He moved his thumb across the mouse pad, but it had no effect. After a short while, a JPEG image was saved to the desktop, and although he could not tell for sure because the commands had streamed by so quickly across the command window, it looked like execution files were downloaded and saved to the computer as well. It was perplexing to him, but he was at the computer's mercy, so he allowed it to continue doing what it was doing. Those were his instructions.

* * *

Angstrom was in his apartment when he heard his computer make a sound to indicate that something had occurred. He walked over to it and saw that a window had been opened on his desktop, and text was appearing within it. Gordon's access to the Ukranian web site prompted a new message to be sent to

Angstrom's PC.

He could tell that a lot of activity was occurring on it; he saw that another image had been received because it flashed momentarily in the background behind the opened window. He didn't try to do anything with the computer while this activity occurred, because he knew by then that he had temporarily lost control of it anyway. It was like a raft going down a river: it moved along the water on its own, and all one could do was sit back and enjoy the ride.

No images were saved onto his desktop, nor any files. From what he could tell (it all flashed by so quickly), an image was temporarily downloaded for extraction of a steganographically embedded message, and as soon as that happened, the carrier was deleted, only to be replaced by a different image, and the process repeated itself many times over. It seemed like a different process this time. He theorized that the encoding density had been significantly decreased, such that the embedded information in each of the steganographically modified images was being hidden deeper and deeper into the image, with less information contained in each one, as an extra precautionary measure to minimize the probability of intercept. All of the extractions would then have to be assembled to create the final message. That was the only idea he could think of as to what was happening. What that meant, then, was that the Professor was being more cautious and had serious concerns about his transmissions being discovered; more concern than when the Candy Mav image was originally used to carry the PrC submission. Otherwise, there would not have been the need to use so many carrier images to send the covert communication payload.

He tried to fathom what might have caused the need for such extra precaution. The images themselves were different, too. They were generic pictures of inanimate objects—chairs, bottles on a table, landscapes. It was as if the manner of the information transmission itself—the multiplicity of modified carrier images with decreased encoding density, and the generic nature of those carrier images—was meant to intrinsically convey a certain message: the situation had changed, the danger level increased. He had no way of knowing for sure, but it was his job to consider such discrepancies; every minor change to what was expected merited serious analysis, all potential ramifications scrutinized.

There was no way for him to know that the Professor was creating the appearance of hiding certain embedded information in the bland images, in accordance with what Dmitri and the Professor had agreed upon, only to be further embedding additional information, known only to the Professor, within those same images. All of the extra levels of steganographic overlay, with the additional, covert information, were accomplished through asynchronous, automated transmissions via a circuitous route of servers and spaghetti code that the Professor had created.

There was one thing that had been clearly communicated to Angstrom when all of the processing was complete: passage for *two* was to be arranged.

So only two were going to defect, Angstrom thought, and he considered the fact that he was not going to get the whole PrC team. Two were better than none, he supposed, especially if one of them was the brilliant scientist that led the whole effort.

The two people referred to in the transmission were the Professor, and one of his colleagues who would be identified at a later date. No name was provided for that second person, which Angstrom thought was odd, but by that time, with all that had transpired thus far, he was not about to try to figure out why the identity of the second person could not yet be disclosed.

What *was* provided, however, was another image of someone. Not the second defector, but someone else—a third person. Angstrom studied the image displayed on his monitor. It was a head-shot of Dmitri in full color, which Angstrom had no way of knowing. He committed the image to memory: deep black hair parted in the middle of his head, long bangs pulled to the sides behind each ear and at shoulder length, and the straight, large, puffy nose.

Instructions were provided as to what was expected to happen next, and one special condition was included, which troubled him and caused him great consternation: the condition of murder. More precisely, assassination.

As a condition to his defection, and as part of the overall plan, the Professor required that a certain person of the KGB, the third person indicated in the transmission, was to be "taken out." The Professor identified that person as one Dmitri Alexandrov. He had changed Dmitri's last name. Otherwise, the Professor knew that his new friends in the United States would search and find Dmitri's real name, and would see that he was not really with the KGB.

Identifying Dmitri as the KGB and requiring his assassination was a calculated risk on the Professor's part. His thinking was that even if the United States indicated its agreement to terminate the man, he knew that it would be a lie. As valuable as the technology was that he dangled in front of them, he knew that it was unlikely that they would go out on a limb and agree to dispose of a KGB agent in exchange for a person that they had never even met. Therefore, the Professor decided to do the next best thing; utilize the American agent in St. Petersburg in such a way so as to at least increase the chances of the Professor making it into the United States safely. Part of that utilization would be to deflect the intentions of someone who might otherwise ultimately try to obstruct the Professor's efforts; that someone being Dmitri. The Professor counted on his request for assassination translating into a mere detention of the target.

Beak-nose was another story—he would certainly do all that *he* could to obstruct the Professor's efforts, but the Professor was not interested in having someone just run interference with that man. He was evil incarnate, and the Professor's escape from Russia would be a hollow victory without also removing from the world the one person that had caused so much harm to him and his wife. No, the Professor had other plans for Beak-nose; he would deal with him directly.

Without even pursuing the matter, Angstrom knew that his organization's answer to the Professor's condition would be "no." Assassination was something that Senior Administrators in

Angstrom's organization rarely condoned, and he knew that they would likely not approve of such an action in relation to two potential civilian defectors. Besides, even though the Professor worded it as such, Angstrom did not really believe the proposed assassination was a mandatory condition; the more he thought about it, the more he came to view it as a request. The Professor wanted out, and based upon the things that Racal said, he *really* wanted out. Even if Angstrom responded in the negative to the Professor's assassination request, the Professor would still want to proceed with the plan to get out of the country; Angstrom could feel it.

He considered whether they should tell the Professor "no" outright, or whether he and Gordon should play along and act like they agreed to the request, only to renege on the promise later, after the Professor was safely in their hands. But perhaps the Professor felt that the KGB man was an obstacle to his defection, following his activities so closely that there was no way the Professor could break free if the KGB agent was not at least *inconvenienced* (which is exactly the line of thought the Professor was hoping for). Angstrom would have to consider the request very carefully before bringing the matter to the Director's attention.

* * *

It was five-twenty-eight in the evening on a Saturday, one week after Gordon had arrived in Russia, and Angstrom was in a communication chamber at CIA headquarters. The small room was filled with various types of communication instrumentation. The particular one that Angstrom was going to use was an encrypted satellite radio link. He looked at his watch to see if it was five-thirty exactly, the time corresponding to two-thirty Sunday morning in St. Petersburg. The time was purposefully chosen, again on a Sunday, with the intention of reducing the probability of intercept. When it was time, he locked onto the appropriate channel to register his radio unit on the system. The radio resource was precious, and each user was supposed to be on the system the least amount of time possible. He initiated a page transmission by depressing a button and waited for the display of the instrumentation to indicate that the called party had answered.

In a small, second floor room of the safe house on the other side of the world, Gordon took out a plastic cylinder that was only fifteen centimeters tall. It was a compact unit—one of the few possessions that he carried with him during his covert entry into the country. He stood the unit on its end, right on the window ledge to increase the received signal strength, slid a chair over to the window, and powered on the unit. The plug of his earpiece was inserted into the jack on the unit, and there he sat, looking out into the dark, early morning sky, waiting for the tone indicating registration on the sat-link. That plastic cylinder housed an entire satellite radio unit; it would have been even smaller if not for the complex, helical antenna system lining its inner wall to increase the electromagnetic directivity for receiving a weak satellite signal.

A few seconds after Angstrom saw that Gordon was registered on the system, Angstrom said into his microphone, "This is Applecore."

In response, and after a slight delay, Angstrom heard through the desktop speaker, in a somewhat garbled voice, "F-bomb's not going to like this." It was a reference to Franklin.

Even under pressure, Gordon still had his sense of humor, Angstrom thought to himself; he envisioned him smiling on the other end of the link. F-bomb was obviously not Franklin's official sat-link codename, but Angstrom knew who Gordon meant. Alone in the safe house, with little food or other conveniences, Gordon still had perspective. That was what Angstrom always liked about him. He was the perfect balance to Angstrom's more serious nature, just like Racal was. They partnered on many missions in the past; sometimes by choice, sometimes because the two of them were the only option, given the gravity of particular situations.

When Gordon heard no response, and knowing they were to be as succinct as possible (their link was of the highest encrypted OFDM signal, but they still had to follow procedure and minimize airtime), he simply asked, "Approval?"

He was referring to the Professor's assassination condition.

Angstrom and Gordon had not communicated anything about it before, but they were both experienced enough to know that amongst the whole plethora of information that the Professor had provided to each of them, including to Gordon when he accessed the site, the assassination condition stood out from all the rest.

The severe frequency-hopping protocol associated with the communication link, compounded by the coarse vocoding and narrow bandwidth, caused the transmitted voice to sound robotic and garbled, but after all of the training he had on the system, Angstrom could still understand it.

"No," he responded.

Gordon would have liked to have paused and weighed the answer, but time on the link was a premium, so he had prepared in advance different lines of communication depending upon the answer that he got.

"Feign agreement?"

Angstrom knew that Gordon was asking the question that he himself had considered: Should they lie to the Professor and indicate that they agreed to the assassination anyway?

"No...communicate no. Next comm: cycle minus D-two. Out.'

That was it. They would not communicate again until the next designated date and time, the minus two indicating that such date was to be moved up by two days.

Gordon had the answer he needed.

* * *

The Professor's instructions for Gordon were provided in a manner that was similar to the one used with Angstrom (with the additional complexity of the layered data): the laptop computer had been taken over, and the instructions were loaded onto it. The whole message delivery was automated in that all of the information was transmitted from a string of remotely located servers, triggered off of Gordon's access to the Ukrainian site.

Knowing that the walls were closing in on him, and that his internet access would be curtailed, if not completely cut off, the Professor's instructions to Gordon contemplated the entirety of the mission, all the way to its end when the Professor and his unidentified colleague were safely out of Russia. Therefore, timing would now be critical; everything needed to occur within predetermined windows of time so that each side could know that the other side's responsibilities were satisfied.

That transmission from the Professor to Gordon was to be the last electronic communication between them. Now, all subsequent communication would have to be through other means. To provide the answer to the Professor's proposed condition, Gordon would have to place an ad in the classified section of a certain edition of a St. Petersburg newspaper. The wording would have to be specifically crafted for the 'positive' or 'negative' response.

Although Gordon spoke Russian fluently, he did not want to place the ad himself for a lot of reasons, not the least of which was his fear that his American accent might arouse suspicion. Therefore, he wrote down what he wanted the ad to say, walked to a small, local business in a strip mall some blocks away from the safe house, and paid an employee in one of the shops to call the ad in for him. The merchant who did him the favor was only too happy to oblige, detecting immediately from the few words spoken as Gordon handed over the handwritten piece of paper, that Gordon was not Russian.

The Professor knew that with his internet access terminated, and Beak-nose's surveillance of him becoming even tighter, that he would not be able to place his own ad in the paper in order to communicate back to Gordon. But there was one more communication that he had to have with him: the indication that he and his unidentified colleague would still defect after receiving the expected negative answer from Gordon. Of course the Professor counted on Gordon's answer being no, but he had to create the perception that he might not proceed if he received such a negative response.

Thus, the Professor had already taken steps to be able to respond to Gordon's answer when it appeared in the form of a classified ad in the newspaper. One of The Mighty Five had been instructed to constantly monitor the daily newspaper for a certain type of classified ad, and upon seeing that ad, he was to mail a postcard, with a predetermined message typed onto it, to a designated Post Office box. It was necessary for the postcard to be sent only after Gordon's ad was placed so that Gordon would see the dated postmark on the postcard and perceive the Professor's answer to be a considered response.

After Gordon had the ad placed with his negative indication, he left the local store, took a cab to another part of the city, got out, and walked a good distance until he arrived at Birch forest in Sosnovka Park, where the Professor had directed him to go. He entered the park at a main path and walked some distance through the wet, muddy ground until he was sure that he was out of site from anyone, and then he strayed off of the path, past many fir trees, into the thick, forest growth. He needed to make his way fairly deeply into it before he was completely out of site, because the foliage had not yet started to sprout leaves so early in the spring season, and it therefore did not provide optimum cover.

The cool air, mixed with the scent of the evergreen trees, was refreshing to him, and he took some deep breadths to enjoy the smells to their fullest. He continued to look toward the designated area off in the distance, as best he could, fighting thorned branches and thistle as he moved, and eventually he recognized some of the landmarks that he was looking for. It was a very laborious effort that he had to go through just to get one Post Office box key, and he was glad to find it encased in the grey metal box when he lifted and pushed away the heavy, dead log for which he was told it would be lying under.

With the key safely in hand, he carefully made his way back out of the forest, making sure that no one saw him as he exited from it, and walked several blocks further to make sure that no one was following him before he finally hailed a cab to get dropped off near the safe house.

It would take a few days for the ad to appear, and the subsequent response to arrive at the Post Office box.

The Professor knew that, and it was done so with purpose when he thought to include the step in the first place. He liked the psychological effect that the wait would have on the person having to check the Post Office box each day to see if the postcard was there.

Therefore, now came the laborious part of the process for Gordon: the waiting. He was well versed in the practice, and he was glad that he had found some English language reading material at the Helsinki station to take with him across the border into Russia. It was going to be a long few days, he thought to himself, and in the dark, upstairs attic with no windows, he pulled on the metal chain to turn on the light, and opened up his copy of Turgenev's *Fathers and Sons*.

In the early evening, but at a different time each day, Gordon took a different route to the Post Office to check for the Professor's response. On the fifth day, when he turned the key and opened the box, he saw a small postcard inside of it. He checked the postmark, and it was dated two days ago, the place of origin being Moscow, with no return address. On the other side, there was only a single, typewritten Russian word: *Prodolzhit'* [Proceed].

14

Misgivings of the Devil

Dmitri was looking down at the paperwork on his desk, and in his state of intense concentration he did not notice that someone had entered his university office. It was only when the door slammed shut and startled him that he looked up and saw that someone was there. The eyes of his visitor seemed darker than usual, more menacing; the black pupils seemed to take up the whole of his eyes, giving them the appearance of black marbles. The head was cleanly shaven, and his face was long and narrow, with a nose that extended outward like a beak ready to tear into flesh, with nostrils flaring.

"Raskalnikov!" Dmitri exclaimed.

Raskalnikov said nothing in reply. His hands were behind his back, still resting on the doorknob, and he looked down at Dmitri with a sinister glare, his eyes squinting and his lips pursed as though he was ready to attack his prey.

"What are you doing here? If you're seen here we're done," said Dmitri, as he got up quickly to go lock the door.

Raskalnikov barely moved out of the way to allow Dmitri

access to the lock. Because he was dressed more formally than usual, Dmitri noticed Raskalnikov's attire: black turtleneck underneath his overcoat, pleated tan slacks, black, leather belt with a silver buckle, and black, leather shoes. In any other setting, it might have been thought of as a progressive-looking business outfit. But there in Dmitri's office at that moment, and given the fact that it was so different than Raskalnikov's normal clothes, it struck a menacing tone.

When Raskalnikov finally responded, the sound of his voice was shrill and rattled Dmitri: "I fear your *Professor* is up to something."

Dmitri could never get used to his voice and the piercing wickedness that it seemed to reflect. Raskalnikov's presence always intimidated him and made him uneasy. He tried to contemplate what Raskalnikov meant by his statement, because the man was often obtuse with him when he spoke, and the fact that Dmitri was so surprised about his unexpected visit made it hard for him to concentrate on what was said. And for Raskalnikov, it was a game he liked to play: he purposefully spoke in vague, unclear terms, and then he would sit back and watch Dmitri squirm and struggle to interpret what Raskalnikov may have meant, full well knowing that Dmitri was afraid to ask.

Dmitri tried to remain calm and look confident after he realized what Raskalnikov meant, and he said, "Of course he's up to something. He thinks he's planning our defection." It didn't work; Dmitri did not look or sound the least bit confident as he spoke.

Raskalnikov ignored his comment and instead pointed to a cabinet in the office, with an inquiring look on his face.

"May I?"

Dmitri, still unnerved by Raskalnikov's presence, at first looked blankly at him as he tried to comprehend yet again what Raskalnikov was referring to. When he realized that Raskalnikov was asking for a drink, he responded, "Yes...please, help yourself. I'll join you." He held out his hand and gestured toward the cabinet.

Raskalnikov opened the cabinet, took a bottle of vodka out of it, and poured them each a glass. He knew that Dmitri was watching him, reeling in fear and anxiety. "Relax, Dmitri. Your friend will not see me here with you. He's not the in the building, I can assure you."

Before handing one of the filled glasses to Dmitri, Raskalnikov picked one of them up, bobbed his head back, and downed the Vodka quickly before re-filling his glass. He then handed one to Dmitri, carried his own refilled glass across the room, and sat down on a couch, with legs crossed and one arm stretched across the back of the couch. While Dmitri took a slow sip of the vodka, Raskalnikov grinned at the sight. He knew Dmitri was terrified, and it was quite enjoyable for him to observe and laugh at his puppet. He continued to watch in silence as Dmitri finished his drink and then poured himself another. Raskalnikov let quite some time pass so that the vodka could begin to have its effect on Dmitri, and calm his nerves.

Raskalnikov said, "I've never asked you before, Pavlovitch, but I'm curious..." There was an intentional pause. Dmitri, still standing up in the office, had been looking down at the floor in silent reserve as he nursed his drink. When Raskalnikov began to speak, Dmitri turned to look at him with an expectant look, waiting for the rest of the inquiry. "Besides the women we give you, and the sums of money, why are you doing it? Why do you betray your closest friend...stab him in the back? Is it the promotion here at the University that you expect to get?"

Dmitri turned his back on the man and walked toward a wall on the other side of the room to where his diplomas and other personal awards were hung. He did not appreciate the highly intrusive question, especially because he knew that it was asked not so much out of curiosity, but to just needle him. He became aggravated, and told himself that it should be enough for Raskalnikov that he was conspiring with him in the first place. To hell with the reason.

Still, if Dmitri were honest, it was a question that he asked himself over and over again, even after he made the decision to do it—to jump off the cliff, as it were. He of course realized that what he was doing would mean the end for the Professor—that man who was not only a brilliant scientist...but who was also a most cultured man; a man who rose above the mundane and the immediate in life, with aspirations for a higher plane of existence in humanity. It was the Professor, he conceded, and his wife (that humble woman), who essentially raised him; rescued him from abusive parents at a very tender age, educated him, nurtured him, instilled culture in him, and treated him like the son they never had. Dmitri knew that in truth it was the Professor who was even responsible for his obtaining the very position he held at the University.

How then, Dmitri would ask himself in those rare moments of self-reflection—when he would contemplate all that he was doing and experience a sense of doubt—could he turn his back on someone like that; obliterate the man, destroy him, he who had done so much for him? It was a question of which he could not allow himself to ever reach a truthful conclusion. He kept that truth buried, deep within his subconscious—his mind involuntarily suppressed it in full recognition of the fact that comprehension of it would result in self-destruction. For, Dmitri's soul was corrupt, and who would want to come to terms with that? It would result in his recognition of the need for reconciliation, and he was not prepared for that. Or rather, he was not prepared for the ancillary result of such repentance: forgoing forever his dreams of climbing the social ladder, of gaining power, and fame.

A fair amount of time had passed in total silence without Dmitri providing any kind of a response to Raskalnikov's probing question. Raskalnikov was enjoying the moment, prying as he was into the evil nature of his co-conspirator, his tool, and shining a spotlight on the man's misdeeds and betrayal—on his tortured soul. He watched Dmitri, studied him, as though he could see the mind contorting itself and plunging into the depths as it struggled to resolve the tensions within it, the outer shell of the body reacting in concert, with facial grimaces and nervous shaking, as if a fist were clenching tightly around the heart and cutting off the very circulation of life.

Dmitri noticed the smirk on Raskalnikov's face, and anger welled up within him. He did not appreciate being studied, reflected upon, and observed, like he was some specimen under the microscope whose every physical detail was noted with curiosity. Nor did he like the spotlight trained upon his iniquity.

"And what about you, Raskalnikov? What are *you* getting out of all of this? Why does it give you so much pleasure…to destroy a man? In service of *Mother Country*, I suppose?"

"Hmph. Don't be ridiculous. I'm not so narrow-minded as that."

He was though. There were no broad, philosophical notions behind the motivations of Raskalnikov. He was an uninspired man engaged in the services of the KGB, and he liked the small amount of power the role afforded him over the weaker citizens at large, the intrigue associated with the position, and the unfounded belief that his efforts would lead to greater things in his career—advancement within the organization. In reality he was a small player in the lower ranks of the KGB, at times a nuisance to it, and those rare occasions of lucidness regarding the reality of his situation only caused an amplification of the wicked tendencies within him, thereby compelling him to direct even more terror against his primary case and subject, his personal punching bag, the Professor.

Dmitri sat back down in his desk chair and turned to face Raskalnikov, who was still sitting on the couch. He wanted to change the subject; the atmosphere seemed to be getting even more uncomfortable for him than it already was, and he knew it was a mistake to verbally challenge Raskalnikov's motives. "The Professor is not up to anything," he finally said, abruptly changing the subject. "He's following the plan, just like he's supposed to."

"But that's just it. My sense is that he's *not* following *the* plan, you imbecile; he's following his *own* plan."

Dmitri, as usual, ignored the slight and said, "Ridiculous. I've known him just about my whole life. If he was up to something, he wouldn't be able to hide it from me." Dmitri spoke with as much conviction as he could muster, but deep down he had begun to fear the same thing himself. The strange non-event at the café was what finally caused that fear to come to the forefront.

The smirk disappeared from Raskalnikov's face, and it was replaced by a visage of controlled power. "He's doing strange things. Our grip is tightening on him by the day, and he's not cracking. In fact, quite the opposite seems true. He's closing in on himself with an apparent sense of confidence and contentedness. And yet again, I'm conflicted...because part of me wonders if he *has* cracked...that he's going to just...blow his head off."

Raskalnikov leaned forward to put his glass on a nearby table, went over to Dmitri, and leaned over him, placing both hands on the armrests of Dmitri's chair. "Another part of me thinks he's up to something—that his little defection plan with you is a ruse and that he's really going to do something else." Raskalnikov's face was so close to Dmitri's that Dmitri could smell the vodka on his breath. Raskalnikov lifted one of his hands off of the armrest and grabbed Dmitri's lower jaw, slowly moving Dmitri's face back and forth. "My friend, if he does—if he screws us over—you'll regret it for the rest of your life. I'll make your life a fucking hell."

Raskalnikov lingered over him for a moment longer, peering into Dmitri's eyes so that the gravity of what he said would sink in and be fully appreciated. He was evil incarnate—the devil personified. Dmitri was terrified.

Mustering what little courage he could, he moved his arm sideways to slowly push Raskalnikov out of the way, and then he stood up. To make an excuse for doing so, he went over to the cabinet to pour himself another drink, and to refill Raskalnikov's glass. "Nothing is going to happen, except for what we've planned to happen," he said as he filled Raskalnikov's glass. "You'll have his head on a silver platter soon enough, available for the offering to your superiors—his and the three others, the rest of his *Mighty Five*." Dmitri said those words in a sarcastic tone in an effort to appeal to Raskalnikov's hatred and loathing of the

Professor, as well as to appeal to his notions of self-importance. "In the mean time, you've got to get out of here; I can't be seen with you."

Raskalnikov watched Dmitri walk to the door, unlock it, and hold it open for him, beckoning him to leave. Raskalnikov, without saying anything further, finished his drink and then walked toward the opened door, seemingly in acceptance of Dmitri's invitation. But before he walked out, he put a hand on Dmitri's shoulder and gave it a squeeze, glaring into his eyes one last time.

* * *

Monday seemed like Saturday...or Sunday...or any other day. With the semester over, his final semester, there were no classes to teach; the Professor could wake up early or sleep late, it did not matter. There was nothing scheduled for him; he had been shut off from the world. Every day felt the same. He wondered how long it would take if he was not doing what he was doing—enacting his plan for escape—before he just shriveled up and withered away. He had no income, and virtually nothing saved. His colleagues would not be able to give him money forever. Play by the rules or die, that was essentially the KGB's ultimatum, and in their minds, the Professor had chosen death.

He got out of bed, put on some lounging clothes, and stepped out into the cold to walk down the driveway. The real signs of spring had not yet been revealed in St. Petersburg, and he breathed in a heavy dose of the fresh air. It was a slow trudge down to the end of the long, gravel driveway, and he pulled his robe closed to keep in the warmth. It was strange to him to be moving about outside that morning, separated as he now was from the daily routine of life. It was like he did not exist. He grabbed the newspaper from its delivery box with anxious anticipation, and then stopped for a moment to look around at the homes sparsely dotting the countryside. Smoke rose from the chimneys of a few of them, the only signs of life so early in the morning. It was moist outside. The smell of the wet earth permeated the air; it felt fresh and seemed pure to him—the land.

On his way back into the house, he grabbed a couple of logs from the wood pile. He threw them into the sleepy embers of the fireplace in his den before sitting in his favorite chair. The only sounds in the house were the crackling of the fresh wood burning, as well as, just at that moment, a single chime from the grandfather clock to indicate the half-hour. When the chime ended, the ticking seconds of the clock fought with and overtook the quietness in the room. The section of the newspaper that he usually never read was the first thing that he searched for and referred to that morning. His eyes traversed the columns of ads until he found the specific one that he was expecting, and there it was: an ad for the sale of a Black Russian Terrier called *"Otritsatel'nyy"* [Negative].

Price: zero. No contact information provided.

He smiled, first because he always knew that the answer to his assassination condition would be "no," and second because just seeing Gordon's ad with the answer in it was a matter of seeing his plan further unfold before his eyes. It was then that the Professor's friend, in response to Gordon's ad, would send the postcard that Gordon would find at the Post Office box with the direction to proceed.

Designing a new military weapon, or planning for one of the courses that the Professor would teach, those kinds of things the Professor had done countless times. But to plan for his defection, with all of the wildcards involved, and the tight surveillance that was upon him, would be his crowning achievement, and to see each step of it successfully transpire gave him a tremendous sense of accomplishment. If he were truly honest with himself, he would even admit that it gave him a thrill. The high probability of something going wrong at so many different junctures along the way, and the fact that so far nothing did, truly energized him; he could feel his goal getting closer and closer, to within his grasp.

The crackle of the fresh wood caused him to look up at it

momentarily. No, he was not surprised that they refused his condition, and he appreciated their honesty.

Now he was at another critical point in the plan—the end game.

15

Meeting, and Infliction of Pain

Gordon did not understand the purpose of what he was supposed to do next. It did not seem to fit within the rest of his instructions, and that caused him to be suspicious. He would be out there in the public, in plain view, vulnerable, just for the purpose of observing something. A discussion with Angstrom would have been greatly appreciated, but radio contact was over. Not until the Professor and he were safely across the border would electronic communication be possible again.

Back when he originally received the Professor's instructions on that PC up in the attic, he studied them intently so that he would understand each step along the way and be able to perform his responsibilities flawlessly. But every time he arrived at this point in the instructions, the part which was about to commence, it puzzled him, and he repeatedly considered what its purpose might be. The most apparent reason for it was the worst of all possibilities: it was a trap, an elaborate artifice to coax a United States agent out of deep cover in order to terminate him. Every time his analysis took him to that scenario, however, he ruled it out, because if that was the goal, there were so many other ways, simpler and more direct, to accomplish it. He could have been ambushed in the forest when he was searching for the key, or picked off while on his way to the Post Office, or even during his return to the safe house. That couldn't be the reason for what he was about to do. He never could convince himself of any plausible reason for this next step, and hence, he was filled with trepidation.

The maze of side streets, marked only by signs written in Cyrillic characters, nearly caused him to get lost on his way to the restaurant called *Na Zdrovye!* It was a Saturday night, on the eve of a performance at a nearby theater, the busiest time to be at the restaurant. He provided his name as Alexondrovas to the man responsible for seating customers, and he was surprised to be seated immediately in accordance with a reservation that was made on his behalf.

It was a prime table, right by the front window facing the street. If not for the present circumstances, it could have been an enjoyable experience for him, dining on fine food and watching the passersby as they walked outside in front of the large window. There was a fair amount of noise in the restaurant from the clanking of glasses against plates, customers conversing at their tables, and chefs feverishly preparing meals in the kitchen. He ordered a bowl of borscht, and the waiter brought it to him piping hot; the beet-red broth was spicy, with strips of lean, slow-cooked beef draped across the top of it.

As far as he could tell as he surveyed the area in between sips of his soup, there was nothing out of the ordinary happening in the restaurant. He struggled to note anything of particular significance; some pair of eyes that might meet his gaze in unspoken communication. He finished the soup and was forced to order a meal so as to keep the generalness of his appearance in tact; a vodka martini was brought to him in the mean time, just to take the edge off. Everywhere he looked, at every table, no one stood out. In fact, he was the only person in the restaurant that was dining alone.

The martini had three pimento-stuffed olives in it. He figured the least he could do was enjoy himself a little while he waited, but he knew to drink it slowly, and not all of it. He would need to remain completely alert.

The Professor had made sure to have Gordon arrive early, before any of the other guests of honor arrived, both to make sure that Gordon was there at the critical time, as well as to give him the opportunity to be the first to see the rest of the other guests as they arrived that evening. The Professor counted on the fact that whoever the United States sent into Russia for this mission would be an experienced agent, and he hoped a *very* experienced agent, such that his or her power of observation would be great—which was certainly the case with Gordon.

Gordon was just about through with his meal when another person entered the restaurant and captured his attention. The gentleman stood out from the rest of the crowd as he was escorted past the tables, because he was there by himself, seemingly dining alone. It reminded Gordon of one of the reasons that he was so nervous and suspicious about being there in the first place, because he figured that he must have stood out just like that other man by virtue of the fact that he was there unescorted. So that the man would not notice him or get a good look at his face, Gordon ducked his head and turned to look out of the window.

The man was seated fairly far away from Gordon, against the wall on the other side of the room. Gordon had to slightly lean over in his chair in order to follow the man as he was directed to his table. When the man sat down, the view of him was obscured by a dark shadow that fell on him, which was cast by a stairway that rose directly above his table. Although Gordon could not make out his face too well given the shadow, he continued to watch him as inconspicuously as possible to try and make out any distinguishing features of him. The most he could discern was that he was bald, and dressed all in black.

After a while, Gordon noticed someone walk by the front window, enter the restaurant, and then lean over and whisper something into the ear of the man working at the front of the restaurant. The worker nodded his head and allowed the new customer to walk through the restaurant by himself.

Gordon nearly choked on his drink when he recognized who it was, for his visage matched the profile of the black and white photograph that had been provided to him by the Professor as part of the downloaded instructions. It was the spitting image of Karl Marx; it was the Professor himself.

On instinct, Gordon quickly turned to the bald man in the shadows to see if he noticed the Professor as well, but before he could look any further, a woman stood in his line of sight, right in front of Gordon at his table. Gordon's eyes traversed the lady, starting at her waist and up to her face, and upon their eyes meeting, the woman said (in Russian), "I am supposed to join you at your table."

Gordon was disarmed. She was beautiful—dressed elegantly in formal evening attire, but not conspicuously, with large, hooped earrings dangling from her ears. He was not prepared for someone to have approached him right at that moment; certainly not someone so attractive. In the instant that it happened, he had no idea what she was doing there and automatically assumed that it must have been a mistake.

She could see that he was confused, and she whispered, in a hurried fashion, "The Professor has arranged for me to join you."

His reflexes instinctively kicked in, and in a manner so as to not draw too much attention to the situation, he smiled, rose slightly as a courtesy, and gestured for her to be seated at his table, which she did. A field agent of his caliber was always able to think on his feet and improvise, and he tried to act as naturally as he could. But he was not prepared for her beauty, and he had to draw upon his utmost ability in order to remain focused on the matter at hand in light of the new development.

"You can't dine alone or you'll be noticed, and you are not to be noticed tonight," she said. He didn't respond except to nod his head in comprehension. She smiled and tried to act natural despite the circumstances, and then added, "I'm going to order a drink, and then study my menu. You must continue to observe all around you. I was told to tell you that at any particular moment, if you deemed it necessary to leave, you should do so at once, and not think twice about it."

Then she instructed him to wave to get the attention of a waiter so that she could order her drink. After doing so, Gordon turned his attention back to the room at large, and more particularly, in search of the Professor. He noticed immediately that the Professor had joined the bald man at his table. So the step of observing had begun, he thought.

As for the Professor, when he went past the tables in the restaurant and walked directly up to Raskalnikov's table, Raskalnikov's jaw dropped when he saw him approach.

"Good evening," he said to Raskalnikov with a serious expression on his face. "May I join you?"

A scowl appeared on Raskalnikov's face in reaction to the sheer boldness of the Professor. "What the hell do you think you're doing?"

"Well," the Professor began as he seated himself, "I was going to ask you the same question. What brings *you* here? Are you expecting someone?" As though to give emphasis to his inquiry, the Professor turned in his chair and looked over his shoulder at the room at large, like he was looking for whomever Raskalnikov might be meeting.

The Professor had let it be known to Dmitri that he was going to meet someone at that restaurant in furtherance of their plan. It was a fraud similar to the one he had perpetrated at the café, but the Professor did not even bother to make up a reason this time as to the purpose of his meeting; he knew that at that point it did not matter anymore, and that Raskalnikov would show up regardless. And so it was true.

When he first heard of the meeting, Raskalnikov was concerned that it might be another hoax. Now that his suspicion was confirmed, he was incensed. It didn't matter though, he thought to himself with confidence, because this time the outcome would be different. This time he had the Professor right where he wanted him, right there in front of him at his table.

"You have a lot of nerve for someone in such a dire situation, Professor."

"Oh, yes? And what situation is that?" Of course the Professor knew exactly what Raskalnikov meant, but he wanted to further aggravate the man and work him into a furious lather.

"You know very well what I'm talking about. You just don't know when to stop, do you? You're like a child, obstinate 'till the end. You're going to be very sorry that you did this, I can assure you, Professor." Raskalnikov leaned forward across the table and continued, "If you thought you've suffered up until this point, you have no idea what's in store for you now."

The Professor knew that was how Raskalnikov would react to their meeting, and it was what he had counted on.

Then he noticed, out of the corner of his eye, that the headwaiter was escorting another lone diner into the room; it was Dmitri. The Professor pretended not to notice him and watched, with his peripheral vision, to see what would happen.

When Dmitri saw the Professor seated at the same table as Raskalnikov, a look of shock appeared on his face. He turned around immediately and left the restaurant in a state of bewilderment.

The Professor returned his attention to Raskalnikov, who saw everything as well, and could see that Raskalnikov was infuriated by Dmitri's actions. With a subtle grin on his face, the Professor said, "Hmm, it seems as though your friend lost his appetite."

Raskalnikov was beside himself at being made such a fool, and it was all he could do to restrain himself from reaching across the table and strangling the Professor with his bare hands. Those very hands gripped the edge of the table and squeezed it tightly, with knuckles turning white under the pressure.

"You're a dead man, Professor. You know that, don't you?"

"That's a very brazen thing to say, even for you, Raskalnikov. Aren't you even concerned that I might be wearing a wire? What if I recorded what you just said and made it known publicly?"

The Professor struggled to appear as confident, composed, and self-assured as he could, but it was not easy for him. He was face-to-face with the man that had tormented him for so long and caused him so much pain in his life; the man responsible for the murder of his wife. Feelings of hatred, fear, and anxiety were experienced, all at once. If he was maybe twenty years younger the stress would not have been so great, but being at such an advanced age, it was incredibly taxing on his nerves.

The thought that the Professor might be wearing a wire had crossed Raskalnikov's mind before he said what he did, but he already decided to himself that he did not care, because the Professor was going to be dead by the end of that evening anyway. Even so, Raskalnikov decided at that moment that he was not going to say anything else to him. He would humor the man no more, and he sat back in his chair, took a sip of his drink, crossed his arms over his chest, and just stared at the Professor. He was going to wait it out; wait for the Professor to get up, leave his table, and exit the restaurant. It was at that point, Raskalnikov told himself, that he would put an end to it all. He had previously hoped that he would have been able to catch the Professor in the actual act of trying to flee the country—it would have looked better to his superiors—but there was no time to wait for that anymore; not after the Professor had done this. His behavior was becoming too erratic as far as he was concerned, and he did not want to take any more chances. Besides, Raskalnikov figured, he could still fabricate some story regarding the defection aspect in order to justify the killing, and make more out of it than there was. Dmitri would be his star witness to back up his account of everything that happened.

The moment of silence between the two of them was beginning to be too much for the Professor; it was starting to make him feel uncomfortable, and as a result, he decided that it was time for him to act. He began to rise from the table slowly, unsure of what, if anything, Raskalnikov might do. When the Professor had run through the various possibilities in his mind prior to this meeting, he concluded that Raskalnikov would ultimately do nothing in the restaurant itself—not in front of so many witnesses. He knew that he would be followed out, though, and that was what he counted on. Sure enough, as soon as he turned and stepped away from the table to leave, Raskalnikov stood up and followed him.

Gordon saw them both rise and begin to leave, and as the Professor walked by his table, the Professor ever so slightly looked over to make eye contact with him, as if to say, "Follow me." A moment later, the Professor was gone, with Raskalnikov not far behind.

Gordon recalled just then how the woman at his table had said, right at the beginning of their meeting, that Gordon might suddenly feel the need to leave at any moment. It was right then that the woman urgently said, "This is it! It's time for you to leave. He's going to need your protection from that man. Follow them — go!"

Gordon was startled to be directed by the woman while he was still settling on the matter himself. When he comprehended what she had said, he immediately stood up, dropped money on the table for their meals, and just before he was about to walk away, he looked down at the woman and asked, "Who are you?"

"I'm nobody; just hired to perform this little act. Go…hurry."

After the Professor left the restaurant, he began to walk down the sidewalk. They were in the heart of St. Petersburg, and the Professor knew the area well, like the back of his hand; that was good for him, because at his age, he was not going to be able to walk for too long, and he needed to make sure that he could get to his intended destination.

Raskalnikov followed him, keeping a distance of about twenty feet between them. He wanted the Professor to know, without a doubt, that he was there and that there was no escaping him — that as soon as the opportunity presented itself, it would all be over.

After about a block and a half, the Professor was winded. He stopped in his tracks and turned to look behind him. Raskalnikov was still there and stopped walking at the same time. The Professor noted Raskalnikov's presence, and then looked beyond him; he was looking for Gordon, and he saw him about another ten yards back. Gordon had stopped walking as well, and their eyes met.

Raskalnikov mistook the Professor's stopping as a sign of fear or confusion, and he relished the moment. His preoccupation with the Professor caused him to be oblivious to Gordon's presence.

Upon confirming Gordon's being there, the Professor turned back around and began walking again at a deliberate pace. Raskalnikov failed to notice that after about three blocks of walking, the Professor had led them to a side street in a less populated area. With fewer people about, Gordon had to be more careful and hung back even further so as to not be noticed by Raskalnikov.

Then, all of sudden, the Professor turned from the sidewalk and went into an alley. Tall, brick buildings bordered both sides, and it was dark and deserted, with trash strewn all over the place. Raskalnikov thought for a moment that the Professor had made a mistake. Then he wondered whether the Professor had finally given up, and that he was ready to be killed; to be put out of his misery, as it were. Why else would he have allowed himself to be trapped in such a place?

The alley dead-ended into another brick building after about another fifty more yards of walking. Further progression was blocked. When the Professor reached the very end of the alley, he stopped and slowly turned around, his back in close proximity to the brick wall. There he saw Raskalnikov, who went up close to him. The black beady eyes glared maliciously, nostrils flaring, and the nose ready to peck the Professor's eyes out.

"May I provide the service of putting you out of your misery, Professor?" Raskalnikov said with a big smile on his face and with his hand held out in an exaggerated fashion, like he was a waiter at a restaurant offering to provide assistance.

There was no response; only silence as the Professor looked at him while he endeavored to catch his breath and struggled to hide how terrified he was.

Raskalnikov was disappointed that the Professor did not say anything. He wanted to prolong the charade a little longer, and savor the moment. The grin disappeared from his face and was replaced by an expression of pure malevolence. He walked a few more steps toward the Professor, and he thought about pulling out his gun, with silencer attached, from the inside of his coat to put a bullet into the man's head. After he had reached into his coat pocket and had his hand on the gun, the thought suddenly occurred to him that such a death would be too easy for the Professor, too quick. The Professor noticed the hesitation and wondered what might be going on inside of the man's head.

Raskalnikov looked around the area for a moment, and when he spotted an old, grimy rope lying amongst the trash, he went over and picked it up and wrapped it through both of his fists in order to get a good grip. With slow steps, he re-approached the Professor, fully expecting him to recoil in fear, at least to some extent, but still there was no such reaction from him. The Professor just stood there, stone-faced, and waited for Raskalnikov, trying to remain as calm as possible.

There they stood, right in front of each other, almost chest-to-chest. Raskalnikov kept his gaze on the Professor's eyes, and while he did so he slowly reached up to apply the rope against the Professor's neck, leaning his body against the Professor's and pushing him back against the brick wall. When the Professor still did not resist, Raskalnikov slowly began to wrap the rope around the Professor's neck in preparation to choke him.

Meanwhile, Gordon, who was in the alley as well and was about thirty yards back from the two of them, was bewildered by what he saw and could not comprehend why the Professor had led them all there and then was not even resisting the man who was about to kill him. He did not recognize the man who was applying the rope against the Professor's neck; it was certainly not the face of the man that the Professor asked to be assassinated.

As for the Professor, he could not predict with certainty how Gordon would react in the present situation. As the calculation went, if Gordon killed Raskalnikov right then and there, then an ultimate aim of the Professor's would have been accomplished. But he could not be sure that would happen, even in this acute situation. At the other extreme, if Gordon did nothing, then all would be lost and for naught, and the United States would not get its prize. So the Professor counted on the fact that Gordon would have to do at least something to prevent injury to the Professor, and he hoped that at a minimum, Gordon would injure Raskalnikov in the process, and hopefully but good.

The Professor could feel the rope as it pressed against the skin of his neck, and he could smell the fetid breath of Raskalnikov as the man stood close to him and breathed into his face. The Professor looked over Raskalnikov's shoulder, over to where Gordon stood, and saw that Gordon was there. Then he saw, as if in slow motion, Gordon pull out an M9 semiautomatic pistol (equipped with a suppressor to lower its sound signature), kneel down on one knee, and put his left hand under his right in support of the gun. An ensuing *thwirp* sound was the only indication that a shot had been fired. The bullet hit Raskalnikov's right arm.

Raskalnikov jerked suddenly and grabbed his elbow as it shattered with the impact of the bullet, and the rope fell to the ground as it was released. Before Raskalnikov could do anything, Gordon lowered his sight and shot him again, this time shattering Raskalnikov's left ankle, thereby causing him to fall to the ground instantly. Gordon immediately rushed over to the man—the whole time staying out of his field of vision so that Raskalnikov could not turn and see who had shot him—and slammed the butt of his gun onto the back of Raskalnikov's head, dealing him a blow that knocked him out instantly.

Gordon stood there for a few moments, looking down at the body to make sure that Raskalnikov was unconscious. Then he turned to look at the Professor, and in Russian said to him, "What is this all about?"

Gordon was surprised at how old the man looked; older than the profile picture he previously saw of him.

The Professor smiled slightly and responded, in Russian, "You did your job well...thank you. We're almost there; you know what you're to do next...I'll see you then. It's the only way." Then the Professor began to move.

"Wait," Gordon said as he gently grabbed the man's arm. "Are there going to be any more surprises like this? Maybe we should do something else, now that we're together." He was careful so as to not squeeze too hard, fragile as the Professor's arm felt to him after he grabbed it. When he thought about it a bit longer, he was surprised that the Professor was able to walk as far as he did just then, considering how old he looked and how frail he seemed.

"This was no surprise," the Professor responded firmly. "It was all planned." He stood and looked at Gordon for a few seconds longer, purposefully so as to allow his face to register in Gordon's memory. "I trust that you'll be where you're supposed to be next, on the appointed day and time. It's the only chance for this to work as it needs to." After a short pause with no response from Gordon, the Professor looked even more intensely into Gordon's eyes and said, "Trust me."

With that, he stepped over Raskalnikov's body and walked away. Gordon watched him, until finally the man turned the corner and was out of sight.

It was a significant risk for Gordon to be carrying his concealed weapon while in the heart of Russia (it had been planted for him at the safe house), and an even greater risk to have used it. Therefore, when the situation presented itself, Gordon was not about to kill the man he presumed was a KGB agent. But the man was holding a rope against the Professor's neck, and he had to do something without being seen by that man. That was when Gordon instinctively decided that he had to incapacitate him with precisely placed shots, which is what he did by shattering his right elbow and left ankle.

By selecting the left ankle instead of the right, he gave the man a better chance to be on his feet again much quicker than he otherwise would have had (albeit with a cumbersome cast on his leg, and perhaps a few metal pins), because the man would be able to use a crutch with his unharmed left arm to alleviate the pressure on his shattered left ankle as he walked. That would not have been possible if the ankle was shattered on the same side of the shattered elbow. Perhaps that small favor, as the theory went, would be enough to serve Gordon's benefit should he by chance be captured at some point in the future. It was a small compromise, calculated at the instant of the moment.

But he also sensed that there must have been something significant about the man he just shot. Otherwise, why would the Professor have taken such pains to lead them all into the alley like that, and trap himself as he did? It was certainly not the man that the Professor had requested to be assassinated.

Gordon looked down at Raskalnikov lying on the ground, unconscious. He decided that he needed to leave before the man woke up or someone saw them there together. It would have been better to clean the area up a little, but there was no time.

Raskalnikov was left there alone. Blood began to surround his body as it overtook the pavement around him.

16

Doubt, Uncertainty, and Fury

It was an embarrassment for Raskalnikov to be found the way he was in that alley: a worker of a nearby restaurant, who was taking out the trash through a back door leading to the alley, found him there, lying in a pool of blood. The ambulance service immediately notified the police when a firearm was found on Raskalnikov's person, and his identity was determined by the identification in the wallet found in his back pocket. The police, in turn, immediately escalated the issue to the KGB when they learned, after checking their internal database, that he was associated with the KGB.

Raskalnikov's superiors had not paid too much attention to him up until that point. While the Professor and his public speeches against the government had been deemed an annoyance, Raskalnikov, on the other hand, had always been an even bigger nuisance to them. The only reason he had been hired into their network in the first place was because of his association with a distant relative who held a position in the organization. Otherwise, Raskalnikov would have never been admitted to the

KGB.

So the Professor represented somewhat of an appreciated distraction, and the fact that Raskalnikov became fixated on the man was to some extent encouraged by his superiors, because it kept him preoccupied and out of the way. But now that violence had been perpetrated against a KGB agent, even one as nettlesome as Raskalnikov, and with a gun no less, they could ignore the situation no longer. Something had to be done to address the matter, and to contain it.

A man waited while emergency surgery was performed. When it was over, Raskalnikov was wheeled out of surgery, still unconscious, and the man had to wait longer, until the effects of the anesthesia wore off.

Eventually, Raskalnikov's eyes slowly opened; everything was blurry to him at first. When he saw the IV attached to his arm, he realized that he was not in his own bed. He tried to raise his head to look around, but the effects of the anesthesia, combined with the blow to his head from Gordon, yielded a severe headache, and he groaned as he let his head fall back onto the pillow.

The man, who up until that point had been reading a newspaper, heard the groan and looked up to see that Raskalnikov's eyes were opened. He walked over to the bedside and looked down at him. An acute sense of embarrassment was experienced by Raskalnikov as soon as he saw his immediate superior standing over him.

"You idiot. What the hell happened?" the man said sternly.

"I was…confronting the Professor, and then I was clipped from behind." Raskalnikov's voice was hoarse from the oxygen and anesthesia fed to him during the surgery.

"What do you mean 'clipped from behind'? Why didn't you have your gun drawn?"

Raskalnikov was still groggy and disoriented, and he was not able to speak as evasively as usual. "I, um…restaurant… someone…" He did not say anything else and instead reached for his head with the hand of his unencumbered arm, acting like he was in severe pain in hopes of deflecting the current line of questioning. His superior became aggravated, fully aware of Raskalnikov's little ploys, but he had no desire to press the issue. It was as if Raskalnikov were a child for whom the man only had so much patience.

Raskalnikov sensed the aggravation in his superior, and the resultant stress he experienced further revived him—his thoughts became more cogent. "He was ready to move...to defect. He was trying to flee the country."

"The Professor? Are you sure?"

"Yes...er...it's imminent. My informant, who is very reliable and knows the Professor intimately...he said so. It...um...it could be any day now. It was the Professor that...lured me into the alley."

Raskalnikov was sorry he said that last part as soon as he did.

"What do you mean *lured* you? Who was in control of the situation?"

That was what Raskalnikov feared his superior would say, and he did not immediately respond to the question.

"Well, never mind," the man finally said in annoyance. "I'm going to assign another couple of agents to him now. They'll pick him up at his house and bring him in. That'll be the end of it. The Investigative Committee has already been called into Special Session in order to find crimes against him."

"Wait," Raskalnikov said all of a sudden, as he reached across his body and grabbed the man's arm. "I want to be there too. I've put a lot of time into this, and...I need to be there."

"Raskalnikov, you can't even walk!" the man shouted at him.

Raskalnikov remained calm, which was not hard given the aftereffects of the anesthesia. "I spoke to the doctor before he put me under. He said that he's seen people injured like this before who were able to get back on their feet in a matter of...days. It will be painful, but I'll be on my feet with the aid of crutches in no time. Just give me a few days."

The man processed what he had just heard in total frustration; even lying on a hospital bed, Raskalnikov was still a nuisance. Just then the surgeon walked in to check on his patient. He knew that Raskalnikov's superior was in the room, and he acknowledged him as he walked past him to check on Raskalnikov. Raskalnikov's superior walked over to the door in the meantime and waited for the doctor's consultation to be finished. When it was, the two met, and Raskalnikov watched as they spoke with each other and made their way out of the room.

* * *

When the two Russian agents parked their car on the country road, some three-hundred feet before the beginning of the Professor's driveway, they were glad winter was finally over and that spring had arrived. It was still cold, but all of the snow was melted; it would have been a very uncomfortable trek for them over the hilly terrain, and for such a distance, if there was still snow on the ground.

It was the day after Raskalnikov's surgery, and while he was still comfortably warm in his hospital room, beginning his painful rehabilitation, the two agents had been dispatched to take the Professor into custody. With one KGB agent already shot down in the process of apprehending the Professor, it was imperative that the Professor be imprisoned and interrogated; the identity of Raskalnikov's assailant needed to be determined.

Fortunately for the two agents, the area was only sparsely populated with country homes, each resting on several acres of land. That, combined with the fact that it was the break of dawn with no one about, made their task of approaching the home unobserved somewhat easier. Having familiarized themselves with the area in advance, they stayed off of the main road and instead went through the brush bordering its shoulder, until finally they reached the Professor's driveway. In order to ensure that the Professor did not see their approach, they walked past the driveway for about another twenty yards, then crossed into the Professor's land and climbed the boulder-strewn slope up to his house. The land was filled with trees to provide cover. Upon arriving at the rear of the house, they stopped by the old shed to survey the area again, confirming its resemblance to the satellite image that they had studied in preparation for their assault.

In their present location, they were out of the line of sight of any of the neighboring homes, and they took advantage of that fact by pausing and going over any final adjustments that might be necessary for their approach. One of the agents saw some movement far off in the distance, so he grabbed his partner's arm and pointed in the general direction. They both froze and studied the area. When the object moved a little closer, they realized that it was only a wild deer roaming the fields, and they both breathed easier.

One of the agents, crouching low so as to avoid any chance of being spotted, crept from the shed to the side of the Professor's house and slowly rose to look into one of its windows. The curtains were not drawn, and he peered inside; there was no sign of life, causing him to believe that the Professor was still asleep in an upstairs bedroom. He turned to the direction of his partner and motioned for him to move to the back door, where they both met. The old aluminum screen door was unlocked, and one agent held it open while the other pulled out his satchel of hand tools and began quietly picking the main door's lock. When it was done, the two looked at each other in order to confirm and synchronize their next move, and when a slight push of the door did not produce any overt creaking noises, they slowly opened it sufficiently for them to enter. Each of them immediately reached into his jacket and pulled out a noise-suppressed Pistolet Besshumnyy. One led, while the other followed close behind as they silently made their way through the first floor of the house, looking for the Professor. When they reached a set of stairs, one went up to the second floor while the other continued his search downstairs.

After a few minutes, the man upstairs came back to the top of the stairs and looked down, where he met the gaze of his partner, who was already standing there. They each shook their heads to indicate that there was no sign of the Professor; he was not there. They remained there all day in hiding, waiting for him to arrive, but he never did.

* * *

Raskalnikov's misfortune put him in an advantageous position within his division of the KGB, even if it was fleeting. The Professor was missing; he never returned to his house that day, nor did he over the next several days. By that time Raskalnikov was out of the hospital and walking with the aid of a crutch on his left side, albeit under the assistance of medication and in great pain. He was determined to walk and to prove to his superiors that he could do so, for he felt that it was his moment to finally demonstrate his worth to them by tracking down the Professor and taking him out. Raskalnikov's superior had picked him up at the hospital that evening, and they were now sitting in his car at the curb in front of Raskalnikov's apartment complex.

"We'll still get him. I don't know where he is at this exact moment, but I'll find him," Raskalnikov said. "My informant has been in on the Professor's plan from day one, and that plan is still on as far as I'm concerned."

His superior was incredulous with him for deluding himself the way that he was. Something had clearly gone wrong — Raskalnikov had been shot twice. Those shots to his elbow and opposite side ankle were precise, the sign of a professional, which greatly concerned Raskalnikov's superior. If the Professor was defecting to the United States and someone shot a KGB agent in furtherance of that effort, then that meant that an active U.S. agent, or sub-agent of a cooperating country, was on Russian soil in St. Petersburg and engaged in a mission. That was a significant development, and Raskalnikov seemed to be oblivious to that significance.

"Don't fuck this up, Raskalnikov, because if you do, it will be the last thing you ever fuck up. Do you understand me?"

"Of course."

Raskalnikov turned in the car to look at his superior and tried to keep up the appearance of being confident. He struggled to exit from the car, the heavy cast on his left foot noticeably weighing him down, and it was a slow walk to the front lobby of his apartment complex. His superior did not pull away until Raskalnikov had entered the building.

When Raskalnikov reached his apartment unit, he quickly put his crutch down and took out his mobile phone. He heard it ring several times at the other end as he nervously waited for Dmitri to answer. Raskalnikov was breaking protocol by contacting his informant by wireless telephone, but he needed to reach him fast, and his mobile was the closest phone available to him.

When Dmitri's phone rang, he suspected who it was. He had not heard from Raskalnikov for some time, which he was glad about, but he knew he would be hearing from him eventually. First it was the botched meeting at the café, and then the one at the restaurant—he could anticipate how angry Raskalnikov was going to be. He was unaware, however, as to what had happened after he had stormed out of that restaurant, and thus he did not know that Raskalnikov had been shot twice.

It was clear to Dmitri that something was seriously wrong, and that the Professor had cut him out of his plans. It caused Dmitri great concern, and put him in a very compromised position, but he of course didn't relay any of this concern to Raskalnikov when he answered the phone. Raskalnikov's tone did not sound good to Dmitri; he detected anger in it, more so than usual, and that, coupled with the days that had gone by without his having heard from him, concerned Dmitri when, at the end of the short call, Raskalnikov insisted upon their meeting up with each other immediately. The place identified was at a store in a shopping center right next to Raskalnikov's apartment complex.

Although it was close to where Raskalnikov lived, Dmitri had to stand there waiting in the designated store aisle for quite some time before Raskalnikov finally arrived, and when he did, Dmitri was stunned at what he saw. The crippled man with a heavy cast on one arm and another on the opposite foot was in stark contrast with the intimidating, overbearing person that he was accustomed to seeing.

"Raskalnikov, what happened?"

"Shut up, you fuck. How many times have I told you not to say my name in public?" He glared at Dmitri, and looked at him like he was ready to kill him. "What happened is your fucking *Professor*, that's what happened! The meeting at the *Na Zdrovye!* was a trap—an ambush! That's twice now that he's played us for fools. This time, though, he's taken it too far." Raskalnikov impatiently proceeded to explain what had happened.

"That's incredible! It's nothing like the Professor. He wouldn't do something like that; certainly not without consulting me."

That was what Dmitri determined in advance that he would say to Raskalnikov, even before he knew that Raskalnikov had been shot; it was what he rehearsed in anticipation of their meeting. He tried his best not to inflame an already volatile situation when he then volunteered, "I'm going to go see him, immediately."

"You can't," Raskalnikov said.

"What? Why not?"

"I had agents at his house for the last several days, and there's no sign of him. He's gone. Flown the coop."

"What do you mean *gone*? I don't get it. The Professor's not living at his house anymore?"

Dmitri should not have kept asking questions. If he was smart, he would have kept his mouth shut, but he was bewildered at what he was learning. As a result, the more he said, the more aggravated Raskalnikov became. It may have been inevitable that Raskalnikov blew up at Dmitri anyway, regardless of whether Dmitri remained silent or not, because Raskalnikov had a burning desire to take his anger and frustration out on someone, and Dmitri, who in a way was responsible for Raskalnikov's misfortune, was the only option available to him. Raskalnikov, leaning on the crutch that was under his left armpit, curled his left hand into a fist and slowly raised it to Dmitri's throat, and then he pressed the knuckle of his index finger up into the soft, fleshy portion under Dmitri's chin. Dmitri could have easily moved away from the crippled man, but he dared not. A few patrons that had entered that store aisle saw what was happening and quickly turned away and went someplace else in the store out of fear, leaving the two alone.

"I said "gone," you shit. He's gone into hiding…or something. He better not be out of the country already."

Raskalnikov dropped his fist because the pressure of the crutch under his armpit was too much. He studied Dmitri's reaction carefully to see if he could detect anything peculiar or noteworthy about it, for Raskalnikov was not completely sure, especially after everything that had transpired, where Dmitri's allegiance truly lay. He was also becoming less and less convinced that Dmitri even figured into the Professor' s ultimate plan. So above all else, Raskalnikov wanted to walk away from his meeting with Dmitri with a better feeling about where Dmitri stood and what his situation was.

Dmitri knew that he was in a very difficult and dangerous position. First, the man that he was betraying, the Professor, seemed to be in fact betraying him. Second, Raskalnikov, the KGB agent with whom he was collaborating, likely suspected now that Dmitri had lost control of the situation, and that perhaps the Professor was not including him in his plans. And third, Dmitri also heard the doubt in Raskalnikov's voice as to his allegiance to Raskalnikov.

Thus, for everything that Raskalnikov suspected or wondered about Dmitri, Dmitri was cognizant of it all as well, and therefore he knew that not only was it highly important as to *what* he said next, but also *how* he said it.

"Listen, Raskalnikov," Dmitri started to say as he looked intently into Raskalnikov's eyes, "first of all, you should know that without a doubt, I'm working for you; don't have any uncertainty in your mind about that."

"Alright," Raskalnikov said, and then after a pause while

returning Dmitri's gaze, he said, "Continue. I'm listening."

"I'm sure that the Professor is proceeding with his plan. I don't doubt it in the least." Dmitri knew that assurance would not be enough for Raskalnikov; not after what happened in that alley. "I've known him for a long time—was privy to his innermost thoughts and notions for years. He doesn't want to be in this country anymore, and he thinks that the other country—the United States—will magically be the answer to all of his problems. He and I have been discussing this for a long, long time, and planning his escape, our escape, continuously. He wholeheartedly believes that I'm going with him. There's no way that he would change his mind, or all of a sudden cut me out of that plan. It would be too dangerous for him to do that now, at such a late stage. Besides, he views me as his closest confidante—his protégé...his...son. He wouldn't cut me out." After a pause, Dmitri continued, "I'll tell you what it really is...he's probably going a little senile."

Dmitri had just thought of that last part at the spur of the moment. He thought it would go a long way toward appeasing Raskalnikov, broaching the notion that the Professor, Raskalnikov's prey, was suffering a mental breakdown. He was massaging Raskalnikov's ego, and his insatiable hatred of the Professor. Raskalnikov continued to look at him without saying anything, and then he took a deep breath and looked in another direction in the store to consider what had just been said.

"You've pushed him hard, Raskalnikov. His inner spirit is dead, and he's a very frail man, both physically and mentally. These things that are happening—these random, unpredictable things—they're all manifestations of a mind that is cracking under pressure."

Raskalnikov listened. Whether it was true or not, he knew deep down inside that he had no choice anyway other than to go along with what Dmitri was saying, because the Professor was nowhere to be found.

He turned back around to face Dmitri and said, "You better damn well be right, Pavlovitch, or it'll be hell to pay." He was looking firmly into Dmitri's eyes as he delivered that message.

Dmitri knew that it was not an idle threat, for he knew all that Raskalnikov was capable of based upon what the Professor had disclosed to him about his wife. His body experienced an involuntary spasm, and he felt real terror in associating with Raskalnikov. It was at that point that he began to seriously question whether he was doing the right thing; wondering what he was ultimately getting himself into by working with this man that was standing there before him. He had an out of body experience where he felt like he was watching the two of them standing there in the aisle of the store. But it was too late to change course now, he thought, for he also did not know where he stood with the Professor anymore, and he didn't have a good feeling as far as that was concerned either.

"All we can do at this point," he said, "is follow through with what we know to be the plan. As you know, the appointed date has been set for quite some time now, and we're almost at that date. We need to follow through according to the plan, even if the Professor has for some reason gone into hiding. For all we know, you may have finally just scared the shit out of him so much that he felt like he needed to do that."

"Going into hiding is one thing. Having me shot is another."

There was more silence as Raskalnikov waited to see how Dmitri responded. Dmitri was not sure what to say and finally realized that it was a no win situation, so he said nothing.

"Do you have any idea who the Professor may have gotten to shoot me like he did?"

"No. None at all."

"I know it wasn't you; you don't strike me as being able to handle a gun like that." Raskalnikov was baiting him now, pushing him and trying anything he could to get him to spit out some useful information.

"Of course not. And there's no one in our circle of friends who could handle a gun like that either, or who would even dare try something like that."

"What about one of his students?"

"No. Impossible. I know all of his students."

Raskalnikov pursed his lips in frustration. He knew that certain things were out of his control and that there was no way to get them within his control. Therefore, they were going to have to do what Dmitri said: follow through with what they knew to be the plan.

The only difference, Raskalnikov pointed out, was that now he was going to have some backup. With the relatively incapacitated state that he was in, there was no choice. Besides, the Professor was showing tendencies that necessitated it. He was unpredictable now.

Raskalnikov also supposed that his co-conspirator, Dmitri, could not be trusted as well.

The backup Raskalnikov was referring to were his two so-called *Angels of Death*—his two trusted henchman—the same two that broke into the Professor's house those years ago and committed the Deplorable Act—the same two men that staked out the Professor's house over the last several days, waiting for the Professor to return. The next time Raskalnikov ran into the Professor, he was going to be fully prepared for him; he was not going to take any more chances.

Dmitri didn't need to know anything further about those two men. The stories he heard were enough. He was in a precarious situation, and he was concerned about what he had gotten himself into; but there was no way out of it.

* * *

Keplar started to feel guilty about what he was doing, but he couldn't help himself. There was something sick and decrepit that he felt about it, but the physical attraction was too great; she held a power over him. Even the fact that she was so much older than him added to her allure. He wondered whether he was experiencing an Oedipus complex. She flew in more often, and they took long lunches together, switching between her place and his. She seemed to enjoy the seemliness of a messy, unkempt bachelor's apartment, and she frequently stayed overnight at his place. She even convinced him to invest in a bigger and higher quality bed.

Besides, he would say to himself, where was Angstrom anyway? He was nowhere to be found. If Angstrom was not going to work hard anymore, why should he?

Keplar still liked Angstrom, and still respected him, but after he became so scarce around the office, without giving any kind of a warning, leaving the team virtually on its own, that rubbed him the wrong way. He felt abandoned, and it was getting lonely around the office.

And what good timing it was for her. She captured her prey just as he was beginning to experience his feelings of abandonment; as soon as she detected that feeling, she converged on him like a lion smelling raw meat. She initially turned her attention to him as a distraction, after being spurned by Angstrom in her suite that day. When she didn't see Angstrom around the office anymore after that, as far as she was concerned, it was because he was avoiding her; it was a rejection of her overtures— one of the rare rejections that she ever experienced.

She liked how attractive Keplar was with his boyish look, and could see the animation in his face whenever she gave him her attention. His young frame and stamina challenged her physically, and she liked being pushed to her physical limits like she was.

After a while, Angstrom's rejection of her became a distant memory. It didn't matter to her anymore, for she soon learned from Keplar how Angstrom had all but checked out. It was the best news she had gotten in a long time, because now he would not be around anymore pushing his ideas concerning what she referred to as fringe submissions, and so the TTO would no longer be distracted by him.

Back during the time when Angstrom *was* around, and especially after he walked out on her that evening, she did not like the feeling of not having him under her control. It bothered her. She was concerned that he was an uncontrollable entity that could ultimately cause her problems. Not anymore. And if she had to later, she would use what she perceived to be his being AWOL against him; she felt like she had something on him, that she could use against him later if she had to. It was something she would keep in her hip pocket.

Sometimes she even whispered negative thoughts about him into the ear of her new lover. She was planting the mental seeds in his head, in the event that she might later need to draw upon its fruit if she ever found herself embroiled in a political battle with Angstrom. One could never have enough allies, she always thought, and having one who was theoretically the right-hand man of Angstrom could prove extremely useful.

Still, she continued to wonder where Angstrom was. His disappearance was a puzzle to everyone (Seavers did not respond to questions about it), and in a way, it added to his mysterious allure, and, if she were completely honest with herself, to her still-existing attraction to him.

Regardless, she turned her attention to the present, and looked at the man that lay on top of her, pumping between her thighs. So young and fresh—he was all hers. She raised one of her legs and drove her stiletto heel into the flesh of his ass, urging him on in his exertions.

17

Delicate Test for a Potential Protégé

The events taking place overseas were now out of Angstrom's control. All he could do was let things run their course, and as a result, he was able to catch up on matters that he had left unattended to for quite some time. When he arrived early one morning at the TTO, Fred Book was standing outside of the building, smoking a cigarette, just as Angstrom had always found him first thing in the morning. Angstrom had already spotted him from a distance as he approached the building, and he wondered what Book would say to him, if anything, after not seeing him for so long.

"Well, look what the cat dragged in," Book said with an exaggerated smile when Angstrom was within earshot.

"Hello Fred, how are you?" Angstrom responded as he continued his steady approach, with his black, leather briefcase slung over his shoulder. When he reached Book, he stopped to shake hands with him, which caught Book off guard. Book noticed immediately the new spirit in Angstrom, and his new, confident attitude.

"I'm fine. And yourself?" Book said.

"I'm good. Very busy, but good." Angstrom briefly looked around the parking lot area, and then up at the sky, with the warm sun shining down upon them. "Looks like it's going to start getting warm around here for good. Spring has finally arrived."

That Angstrom characterized himself as being very busy, despite being away from the office for so long, interested Book. He tossed his cigarette onto the ground and twisted his foot on it to put it out. "Yeah. Makes smoking a cigarette out here a hell of a lot easier, that's for sure."

"How's the submission review been going? Any news I should know about?"

It was the first time that Angstrom had ever asked Book his opinion about anything, work or otherwise, and Book was surprised by the sudden question, but actually appreciated it. The smallest question can sometimes make all the difference in the world for a person. He proceeded to update Angstrom as much as he could about the project's status.

It was when he said that Keplar had been spending a lot of time with Rand that a noticeable change appeared on Angstrom's face—there was a flash of disappointment, but it was just long enough for it to register with Book. It made Book wonder as to the meaning behind it.

When Angstrom tried to find out more from him as to what Keplar and Rand had been working on, Book claimed ignorance, which was partially true. He did, however, say that he noticed that they seemed to be spending a lot of time together outside of the office as well, and that they took long lunches together. Angstrom could tell, without anything expressly stated, what Book was implying, and Book, in turn, knew that Angstrom had gotten his meaning.

"Alright, thanks for letting me know," Angstrom said as he gently tapped Book on his arm as a gesture of thanks. He started to walk away and toward the building but then stopped. "Fred, I've got a special submission that I'd like you to start looking at. Only this one is different; I want it to be kept a secret. It's not to be discussed with anyone. Can I trust you on that?"

"Um, sure. Yes…you can…only…"

"What? What is it?"

"Well, it's like I just said: the word I got from Seavers was that they were close to identifying a winning baseline submission for Chassis and Armor, and that it would be announced soon."

"That's okay, don't worry about that. I need you to really dig into this new submission that I'm going to give you; every page, and every detail of it." Angstrom had to throw him a bone in order to further ensure his adherence to his request for secrecy, so he said, "I want to know what you think about it. I'll let you know what, if anything, I want you to do with it after that, but I have a feeling that after a certain amount of time studying it, it's going to become obvious to you. Can I ask you to do that?"

The cryptic nature of what Angstrom asked puzzled Book. And there was also that intangible difference about Angstrom; something that Book could not put his finger on, but was there. Angstrom seemed more alive, or engaged. Book could not describe it if he had to, and the powers of observation and interpretation were not his strong suit. But Angstrom's reappearance that day, coupled with the new presence about him, somewhat endeared Angstrom to him, and he therefore wanted to comply with what was asked of him.

"All right, John. I can do that."

"Good. I'm going to stop by your office this morning to give it to you. Again, I can't stress enough the importance of its being kept a secret, even amongst the rest of the team. I want you to study it in your office, like you did the other submissions. Keep it locked up and don't check it back in with me, alright?"

A slight suspicion was aroused, and Book studied him for a moment. "Does Seavers know about this?"

"No, he doesn't. Not yet. But he will, and he'll be pleased when he does." The response was genuine. After having some distance between himself and the TTO, both temporally and physically, he was able to see some things differently. One of them was that he wanted to utilize Book as a conduit for the PrC submission—a conduit to cause an actual, thorough consideration of it by the larger committee.

"Alright then. Just bring it buy my office like you said."

"I will…and thanks."

They shook hands again before Angstrom continued on his way into the building. Book reached into his jacket to pull out his package of cigarettes and grab another smoke.

Angstrom went to his own office and began working on some paperwork that was piled up on his desk. There was a lot to catch up on, and he was glad about that, because it would help keep his mind off things that were happening overseas.

"Andersen, how are you?" Angstrom said as he poked his head into Keplar's office later that morning.

Keplar was surprised to see him for the first time in so long, and a smile, combined with a perplexed expression, appeared on his face. "Hey, long time no see. I'm good…I'm good. Where the hell have you been?"

Angstrom entered the office and had a seat in front of Keplar's desk. "Sorry, I didn't have a chance to tell you when it came up, but I've been off doing some special work for Seavers. It happened suddenly and was urgent. It's not finished yet, but I have a little break, so I thought I'd drop by and touch base with you."

"Hmm. What kind of special work, if you don't mind me asking?"

Keplar was harboring the impression, based upon the seeds that Rand had planted with him, that Angstrom was not up to anything as far as the TTO was concerned, and that actually, Angstrom was in a sense just shirking his duties. He was not, however, completely sold on that notion, because as much as Rand had him under her spell, there were still a few remnants left of an independent-thinking Keplar, and now that Angstrom was back, he wanted to give him the benefit of the doubt.

"Unfortunately, I can't say," Angstrom said. "Maybe I'll be

able to disclose more in the future, but I can't right now." Before Keplar was able to say anything in response, Angstrom held up one of his hands and said, "I know, I know, sounds evasive. But... give it some time." Then he looked around Keplar's office a little and said, "What's been going on around here? What's the status of Chassis and Armor?"

Keplar did not appreciate Angstrom's sudden change of the subject, but he let it go. "Well, things have been moving along. The committee has narrowed the initial Chassis and Armor submissions down to two, one of them coming from our subgroup."

"Oh yeah? Which one?"

Keplar took a large binder from a shelf behind him and slid it across the desk toward Angstrom, who in turn opened it to the inside front page. Without having to look for too long, he recognized the submission.

"Right, I remember this one... from one of the tried and true, so to speak, so I guess I'm not surprised. How did this one get selected without me being around?" He thought about that after he said it, and realized it could have been misconstrued. "I'm not upset or anything; I'm just curious."

"Well, it was Susan. She started making herself more visible after she learned that you weren't...around. So when the time came, she asked about our frontrunners, I told her, and then, between the two of us, we selected this one."

"I see." To himself more so than to Keplar, he said, "So it's this one, and one other one, that will be presented to the Director and the rest of the selection committee." He closed the binder and, still looking down at it, thought about something for a moment.

"What about *vehicleforge.mil*?" he finally said.

"Huh? What about it?"

"It's supposed to be the server system for the crowdsourcing collaboration. Have you guys seen much meaningful activity on it?"

"Sorry for asking, but when you say *you guys*, who do you mean? Are you referring to Book and myself?"

"Alright then, have *you* seen any meaningful activity on it?"

"Well…" Keplar paused and squirmed in his chair a bit before continuing, "to be honest, I haven't really looked at it."

"Why not? I told you that we needed to. Our sub-group had sole responsibility for monitoring it." Angstrom was aggravated, and it showed in his voice.

Angstrom's tone, combined with having been away from the office for so long, did not sit well with Keplar. "John, you weren't here, and I was burning the midnight oil trying to keep things going. I did the best that I could under the circumstances, and I just couldn't get to it." They looked at each other for a moment. Angstrom was unimpressed, and Keplar could tell. "Besides, when Susan started managing things in your absence, she told me to not spin my wheels on vehicleforge and to just focus on our finalists. So I spent a lot of my time reaching out to the submitters to get their last minute modifications to their submissions. Rand said it would be a good networking opportunity for me."

Angstrom did not quite believe what Keplar said, except for maybe the networking part, because before Angstrom withdrew from his activities at the TTO, they were down to a narrow set of finalists already, so Keplar should've had plenty of time to at least visit vehicleforge.mil to see what was there and what type of collaboration was occurring. Not monitoring that facet of the program was a huge oversight on his team's part, and he knew it. There was a lot of untapped material there. Fortunately, and unbeknownst to Keplar, Angstrom suggested to Seavers some time ago, when he knew that he would not be around for a while, that another team should also monitor the site in parallel to his own team's efforts. Without having to say anything explicitly to Seavers, Angstrom could tell by the look that he got from him that he understood.

Angstrom wanted to diffuse the situation with Keplar and avoid his getting more agitated than he already apparently was, and so he said, "I have something for you."

The sudden change of topic had its desired effect; Keplar's body relaxed; his shoulders dropped slightly, he sat back in his chair, and his facial muscles un-tensed themselves. Angstrom took the thin booklet that he was holding under his arm, which had gone unnoticed by Keplar up until that point, and tossed it onto the desk. It was much thinner than the complete version of the PrC submission that was given to Book, and subtle changes were carefully made throughout it—numbers, calculations, and the like—so that Angstrom could later trace certain aspects of it back to Keplar.

Angstrom had considered the fact that Book had been around for a long time, and Seavers seemed to trust him, however ineffective he was in his role. So he felt comfortable giving him the complete PrC submission (minus the Dostoevsky quote and the reference to PrC). The worst that could happen was that Book would disobey Angstrom's instructions and share the submission with Seavers, which would have been a nuisance, but it wouldn't have been the worst thing in the world.

With Keplar, however, there was no such track record, and he had to be more careful with him. And his ties to Rand greatly concerned him as well.

"This is a different submission," Angstrom said as he tapped the booklet with his index finger. "It's one you haven't seen before. In fact, it's not even the entire submission. I'd like you to study it carefully, and then I'll want to talk with you about it, one-on-one. Don't let anyone else see it, don't discuss it with anyone, and don't even let anyone know that you have it or that you're studying it; not Book, not Rand, not even Seavers."

Keplar switched his attention from Angstrom over to the booklet sitting on his desk, and he picked it up and started paging through it. "Hey, some of this looks familiar," he said after a minute. He perused it further. "Yeah, you mentioned some of this stuff to me before. Is this where it came from?" He flipped to the front cover and first few pages to see who the submitter was.

"Yes, that's correct."

"There's no company or anything on it. Who submitted it?"

"All such identification has been removed. For now, that's being kept secret."

"Why's that? And why can't I even discuss it with Susan... or Seavers? That's pretty odd, John."

"Look, I just need you to do this for me. I'm not asking you to break the law, or TTO policy, or anything like that. I just need you to keep this between you and me. Can you do that?"

Keplar looked at him with a queer expression on his face. He was trying to reconcile what Rand had been saying about Angstrom behind his back with what Angstrom was saying to him now. Still somewhat puzzled, he finally said, "Sure, I can do that. No problem."

"Good," Angstrom said. He slapped both hands on the armrests of his chair and stood up. "Obviously you'll want to keep the booklet out of sight and locked up when not in use. Keep it here, in your office, rather than in the data room. No one sees it." He looked at his wristwatch and said, "I've got some other things to take care of, so I've gotta run. I'll see you later."

"Are you back for good?"

Angstrom was already halfway out of the door when he turned to Keplar and replied, "No, I'm not. You'll need to work independently on what I've given you for a while, and we'll go from there."

Keplar raised both of his eyebrows for a moment, indicative of his sense of incredulousness. "Alright, whatever you say."

Angstrom had the data room all to himself. Book and Keplar would each be in their offices studying the PrC copies that Angstrom had given them. He sat down in front of one of the computer terminals and logged into *vehicleforge.mil*.

There were a large number of files and data there, and chat rooms on various topics with active participants. Angstrom navigated to the document repository that he was looking for — Chassis and Armor Submissions — and found many large files there. The cursor ran over each one so as to highlight the file name, and the submitter's identity could be determined for a lot of them just by doing that. For files where that was not the case, he had to open them and go to the first few pages to determine the identity of the submitter. For over two hours, he surveyed all of the submission material on the server, looking for anything that was out of the ordinary. It was a slow process, because there ended up being many files for which the identities could not be determined by scrolling over the file name, and so each of them had to be opened and quickly reviewed. What made the process take even longer was that the files were large and thus took a longer time for the computer to open them. The time necessary to open each file, when considered in the aggregate, was not insignificant. He also had to avoid getting distracted from what he was trying to do, because some of the files disclosed interesting technology, and he found himself tempted to stop at various points to study the information.

When there were only a few files left, his cursor scrolled over one that caught his attention.   The name of the file was *"source*forged.Armor"*. For some reason it seemed curious to him; his intuition told him that there was something unusual about it. He scanned the information associated with the file, which was displayed adjacent to the file name along the same horizontal line, and saw that the size of the file was only about fifteen kilobytes. He opened the document, and after a short title page, there was only one more page. It was a full-page, color image of Candy Mav; the picture that was on the CD from the magazine. He sat back in his chair to look at it for a moment. The DoD's security software passed it through to the vehicleforge server just like all of the other files.   There was a small line of text underneath the Candy Mav image. It was the html link:

*http://www.imagecode/microsite/patternsubfile/index/
CLEV32/reflected.kz*

Seeing the image and the link fascinated him, and he pondered what it meant.   The Professor had initiated several forms of contact with the DoD to increase the likelihood of getting through to someone.   Angstrom marveled at the cleverness and thoroughness of it all. In thinking about it further, he concluded that the redundancy absolutely *was* necessary, because as it was, clearly no one opened up the vehicleforge file, or if it had been opened, no one followed through and went to the web site.

Angstrom was in and out of the TTO over the course of the next couple of weeks. He checked in with Seavers and filed a few reports as required, and he was on good terms with him. Then one late afternoon, while Angstrom was sitting alone in the data room, the door opened, and he heard a female voice say, "Where's my Mr. Kep..."

Just as Rand said those words she saw Angstrom sitting in the room. "Oh, um...John, I didn't know you were here."

She was astonished to see him. When she recovered from her initial shock, she became pleased and sat down in one of the chairs opposite him. As always, she was impeccably dressed. She crossed her bare legs, flaunting them, accentuated as they were by a pair of Christian Louboutin pumps.

"Well, you finally decided to make an appearance," she said. "Where have you been?"

She had inquired of Angstrom's whereabouts with Seavers, but he had always carefully rebuffed her.

"I've been busy taking care of things."

"Hmm," she said as she looked around the room, as if in comment to his statement, because all of the work was in neatly organized piles, essentially complete. "And what might that be? No one's seen you around here in a long time, and Seavers never seems to want to tell me what you're up to."

"Let's just say that I was asked to do some special research on something."

"Right. What kind of special research?" She had a smile on her face; she was being playful, while at the same time somewhat assertive.

Angstrom tried to deflect her questions by asking one of his own. "What have *you* been up to?"

She didn't appreciate his turning the questions around to her, but she played along. "Oh, we've been putting the final touches on selecting the baseline finalists," she said as she held out the palm of her hand and waived it around the room in reference to all of the submission materials. "There's one that came from your team, in case you didn't know."

"Yes, I do know. Andersen told me." He was thankful for what she had just said, because she opened the door to something he really wanted to talk about with her, or rather, to delve into. It was only a slight opening, but he was skilled enough to make use of it nonetheless, and he stayed on the subject. "I guess I shouldn't be surprised about the one selected from my team, but I am disappointed."

"Disappointed? What do you mean disappointed? Why? Your team did an excellent job."

"The technology was not cutting edge enough. It was a conservative submission, the same old thing, just scaled down to meet cost. There were other ones...correct that..." he started to say as he held up an index finger, "*are* other ones that are still worthy of consideration."

"I doubt that."

She picked up a random file that was sitting nearby and looked at its cover. She was trying to decide whether she herself wanted to delve into something in particular or not. She was also still smarting a little from the abrupt way that he had left her the last time they were together, which she was now reminded of after seeing him again. She had not been with him since then, and such separation, along with the passage of time, inevitably caused a mixture of emotions in her. She decided to change the subject—to take a chance.

"Hey, it's Friday, and I'm staying in town over the weekend. Let's get out of here; you want to?" She already had plans with Keplar, but she knew they could be easily cancelled.

It was going to take some work if he was going to get from her what he wanted, and as a result, he knew he was going to have to play along with her for a while, and humor her. "What do you have in mind?" he said.

"How about dinner?"

"Well, the thing is, being away from the office as much as I have been, I've got a lot of catching up to do." He was teasing her; stringing her along for a bit and toying with her emotions. He knew that she really wanted to go out, and so did he, but that little bit of tension caused by his feigned reluctance might serve him well; it would make her more anxious—more unguarded.

"I'll tell you what," he said. "I've been looking forward to catching up with you and your work. How about if we stop at *Le Cezzane's* and do carry out back to my place? You can bring some of your work with you and we can work together; I'm really interested in getting your thoughts on some things." He had thrown several enticements at her all at once, the most prominent one being the offer to go to his place. He knew that, at least at one time, she was dying to see it.

"Hmm. The thought of discussing work on a Friday night doesn't really intrigue me...but I do love Cezzane's." She was, in fact, interested in the offer, but she too wanted to play the game a little, so she hesitated before providing a response. In the end, his instincts were correct; she was still very desirous of seeing where he lived, even after all of the time that had passed. "Alright, let me grab my briefcase, and then we'll go," she finally said.

He wasn't through with his manipulation just yet, however. Reaching into his pocket, he took his keys out, separated the one for his apartment from the rest of them, and handed it to her. "Here."

She leaned over in her chair, her legs still crossed, and took it from him. "What's this?"

"The key to my apartment. I've got an errand to run before I go home, and I'm sure you won't want to go with me. There's some Champagne in the cabinet next to the refrigerator; put it in the freezer to chill, and then make yourself comfortable. As soon as I've run my errand, I'll stop and get the carry out, and then I'll meet you at my place. How does that sound?"

He knew that she would jump at the opportunity to spend

some time alone in his apartment, thereby giving her the opportunity to explore it unhindered. The apartment was a perfect reflection of him, and he knew it. He was counting on the fact that his opening it up to her and letting her explore it would help re-ignite her interest in him that much more.

"Are you sure? I'd be happy to ride with you." In actuality, she was thrilled with the idea of being able to roam freely in his apartment without his being there. She had a strong feeling that what she saw there would give her some insight about the man who was otherwise so distant and mysterious to her. Angstrom's assessment of her was spot-on.

"No, no. It's all right. You go on ahead. Make yourself at home."

Still he was not done. He had to prime the pump one more time, and he said, "Think about what you'd like to update me on about work while you're waiting for me, and then we'll have a good talk about it when I get there. I'll see you in about an hour." He tore a sheet of paper out of a notebook, wrote his address on it, and handed it to her.

She took the slip of paper from him and looked at it, and then at him. Part of her wondered whether he really wanted to talk about work or not. She slowly and purposefully adjusted herself in her chair, and Angstrom could hear the subtle sound of the skin of her bare legs rubbing together as she ever-so-slightly moved them. His eyes instinctively looked down at them, which she noticed. A slight smile appeared on her face. After that, she rose from her chair to leave, and as she did so, he watched her and didn't hide the fact that he was watching her. Her shapely hips moved just a tad more than they usually did.

After the door closed, he dropped his head forward and massaged his temples. Then he slapped his cheek a couple of times. The cat-and-mouse game was one that he had played countless times in his career, but it seemed to be getting harder for him. He was getting too caught up in it; even with all of his experience, it was not easy for him to turn things on and off anymore. Maybe he was getting too old for it, he thought, and the position of handler couldn't have come at a better time.

He shook his head to regroup and try to get her out of his thoughts—he leaned his head back and let it rest on the chair's headrest for a while. When he had sufficiently calmed himself, he went over to Seavers to touch base and pass the time; he was in no hurry to get home, and had nothing else to do before he went there, but he wanted her to be on ice for a while.

* * *

His apartment was just as she imagined it would be: located on one of the top floors of a high-rise building, sparsely furnished, with geometrically influenced appointments throughout, and earth-tone colors interspersed occasionally with a few darker ones, including a brown, heavily-worn leather sofa set in the living room.   When she first got there it was already early evening, and soon the sun fell completely into the horizon; the floor-to-ceiling windows grew dark, and so did the inside of the apartment. She turned on a couple of lamps in the living room and noticed that the place seemed to take on an even more mysterious, masculine ambiance.   Everything about the place breathed Angstrom to her; it was unlike any of the other places of the men she visited. She slid her pumps off in order to move around more comfortably, but placed them in a location where she could easily get to them as soon as Angstrom arrived. She wanted to look her best for him.

A couple of framed photos of Angstrom's daughters—the only photos of any kind in the apartment—rested on a hallway table, and she picked each of them up to study them. One of the girls looked just like him; she could see it. The other, the older girl, looked different, presumably more like his ex-wife, though there was no photo of her with which she could test her assumption. There was a book shelf on the other side of the living room, and she walked over to it, curious to see what he liked to read; there was a diverse range of topics, both fiction and non-fiction. John Le Carre, Charles McCarry, and Len Deighton were well represented from the spy genre.

After walking all through the apartment and exhausting her exploration of it, she made her way back to the bookshelf in the living room, picked something out to pass the time, and curled up in a brown, leather chair that faced one of the large windows that looked out to the city. She was there for quite a while, switching her attention back and forth between the book in her lap and the sparkling lights of the city seen through the window. Eventually she heard Angstrom at the door, but she was so comfortable in the chair that she decided to not bother getting up to put her pumps back on.

"Hello there," he said. "It looks like you've made yourself comfortable. Excellent." He was carrying a large bag of carry-out food. "Don't bother getting up; I'll set the table."

She stood anyway and said, "Let me help you."

As he placed the bag down on the dining room table, he said, "I'll tell you what, you start taking the food out of the bag, and I'll get plates and silverware. Would you like some champagne?"

"That would be great. By the way, I love your apartment. For some reason it really speaks 'John Angstrom.' "

He laughed and said, "Well, good, I guess." He was in the kitchen already.

"Oooh, tapas, I love them," she said as she opened some of the food containers and placed them on the table. "I didn't know you could get this to go from Cezzane's."

"I'm glad you like them," he said from across the room. "I wasn't sure when I got it. I just thought it would be something light and not weigh us down. I'm on pretty good terms with the owner, so he has it done special for me."

"Marvelous, I can't wait," she said as she continued to look at all of the opened containers and smell them.

"So," he said while starting to partake of some of the food after they had sat down. "What are your thoughts on the finalists for the Chassis and Armor effort?"

"Oh come on, John. It's Friday night. You're not serious about wanting to talk about work, are you?" she said as she took some Chopitos out of a container and put it onto her plate.

Angstrom was careful in responding. He needed to guide the conversation to where he wanted it to go, and it was a delicate task. "Maybe just a little," he said. "How about if we get some business out of the way, and then we enjoy the evening after that?"

"You promise?"

"Yes. I do." He watched her as he ate some food, hoping that she would come around.

"All right, fine," she said. "The first phase of the Chassis and Armor crowdsourcing effort is coming to a close, and there are going to be three finalists for the baseline, including the one from your group." She listed the other two tentative finalists from the other sub-committees, and Angstrom was not surprised by them.

"What about any of the University submissions? I thought that there were a couple of efforts in my group that had some aspects that merited consideration."

"Yeah, maybe. Probably not, though."

"What useful information did we get through vehicleforge.mil? I thought that was an interesting concept for the effort." He wanted to know if any of the other Chassis and Armor sub-committees noticed or made any comments on the Candy Mav image that he found on the server.

"Nothing really, so far. It isn't really time for that yet, but it will be. It's going to be used for the Phase Two effort for sure, but honestly, we really didn't put a whole lot of energy into monitoring the activity on it for Phase One."

"That's too bad. I looked at it recently and saw some interesting technology there."

"Really? Like what?" Rand was also taking the opportunity to probe Angstrom. If she could get an inkling about what he might've been secretly working on for so long, so much the better.

"Well, I witnessed some real-time collaboration by several universities critiquing each other's submissions and offering suggestions and improvements for them. It was quite fascinating. I mean, that's the whole point of it, right?"

"In part, sure."

She still didn't offer too much, he noticed. She had a wall up just like he did, and so he would need to go further with her, press her into it, and see what his inquiry would turn up, if anything.

They were each there that evening for their own reasons. For him, it was a suspicion about something in particular, and he needed to find out whether his suspicion was true. He hoped not, but it was extremely important to investigate the issue and determine the truth, because it would have significant, long-term implications, either way. For her, it was not so much the need to control anymore, though she would gladly accept that if it came. She wanted him now; all of him; mind, body, and spirit. She finally realized that fact the minute she saw him again in the data room, and the more they spoke that day, the more that feeling intensified. It was the first time she ever felt that way about a man before.

Angstrom decided to slow down for a moment and just eat his food; if he appeared to press too much, he feared that she might become suspicious and realize that he was fishing for something. She noticed his pause, and the silence soon had its desired effect on her.

"So, are we done talking about work?"

"We haven't really discussed it too much, have we? I sort of feel like we've been dancing around each other for some reason. I'll tell you what. Let's put this line of conversation on hold for a little bit and just enjoy the champagne and the great food. We can always talk about work afterwards."

That suggestion didn't sit well with her at all. She didn't like the prospect of work hanging over their heads for the entire dinner.

If Rand thought that Angstrom seemed different now, Angstrom thought the same of her, too; he just noticed it. She didn't seem as overconfident as she used to be; there was almost a vulnerability about her. The reason for it was not apparent to him, but if any of it was attributable to him, he thought that maybe it was because of the way he left her so abruptly at her suite that one evening. He had forgotten about that until just then. And also, he thought, his not being around the office, and Seavers not telling her why, could've caused her insecurities, or whatever feelings she was experiencing, to fester. He began to develop a little sympathy for her, at least for the time being.

"Like hell we'll talk about work later. Why don't you just come out and tell me what you've been working on so secretly. It's not that thing called the PX Submission, is it?"

And there it was!

She just came out and gave it to him. The PX Submission—that was the name that he had placed on the abbreviated PrC submission that he had given to Keplar. Angstrom had assigned it a different, unique name before giving it to him, both to maintain secrecy about the PrC matter itself, but also to provide it a unique identifier that could be traced back to Keplar.

So Angstrom's suspicion was true: Keplar, despite specific instructions, shared the submission with Rand. Angstrom feared that the young man might've been too far gone with her, caught in her sexual grip, such that he would even disobey Angstrom's direct order. It was disappointing for Angstrom to receive that confirmation—greatly disappointing—not the end of the world, but incredibly unfortunate nonetheless, because an agent not being able to control himself sexually made for a bad agent, and Keplar's sexual appetite was certainly too great and unmanageable. It was the one concern that Angstrom had about Keplar in considering him as a protégé.

At that point in the evening, the challenge was over with Rand; he had gotten what he needed from her. But with the sudden change in how he viewed her, or, said another way, his seeing a more humane side of her, and not just a manipulative, sexual predator, he wanted to be a bit more elegant in making a graceful exit versus the last time that they were together. Despite the fact that he disliked what she represented professionally, and the tactics she used, he didn't want to in essence just kick her out of his apartment. For some reason, he began to wonder about what could've happened in her life to make her turn out the way that she did. Sexual abuse? An overbearing, mentally abusive parent? Or was it just a matter of unbridled ambition with no scruples to temper it? It had to be something. A person doesn't just turn into a sexual piranha for no reason.

Besides, on a more practical, self-preservation level, he hadn't accepted the position of handler yet; it wasn't a done deal, so bridges could not be burned, and Rand definitely represented a bridge.

Therefore, he played along and pretended that her suspicion—that he had been doing a lot of research on the PX Submission—was correct, and that was the reason why he had been out of the office for so long. He carefully massaged her ego a little by asking her a lot of technical questions about it, such as issues related to manufacturability, and potential quality roadblocks that might be encountered.

His approach worked. She appreciated being consulted by him the way that she was, and it made her feel like she was finally in the loop on things—on everything. The discussion continued for close to two hours, and Rand was quite passionate in her discourse of PX. She gained more and more confidence on the subject matter the more Angstrom asked her some purposefully basic questions about it, and the confidence began to give her the feeling that maybe she could even get the advantage back over him, and that maybe there was a chance for the two of them after all.

After dinner, they brought a bottle of wine with them over to the couch and talked for another couple of hours, only this time they talked about each other—their backgrounds, aspirations, and things they liked doing in their free time; anything but work. Angstrom made most of his part up, but he was honest with her in certain areas, where he was able to. He was ultimately unable to get any highly personal information out of her, including what might have been the deep-seated cause for her confused, sexual nature. Even for a manipulator as skilled and as experienced as

Angstrom, one evening was hardly enough time to uncover the depths of one's soul.

When it got very late, Angstrom offered to let her spend the night, and for the briefest of moments she became thrilled with the thought that she had made a breakthrough with him. When he clarified that she could stay in the separate guest bedroom, and based upon how serious and genuine he appeared to be in presenting the offer, she politely declined with disappointment, because she didn't get the feeling that if she accepted the offer that it would lead to anything. Sleeping in a guest room was not her style, and it would've been another blow to her ego, if in fact it did not lead to anything. Her thinking was that it was better to go home and leave the door open for another day, rather than try to force the issue right then and there. There was a newfound prudence in her with respect to her approach with Angstrom.

After he escorted her down to her car, she couldn't help but think, on her drive home, that something was not quite right about the evening that they just had together. He was different in the way that he interacted with her; at times he almost seemed gentle and compassionate, and she wondered if it was all genuine. She lay awake half the night thinking about it.

On his way back up to his apartment, he thought about what he had learned that evening about Keplar and the made-up PX submission. His mind was still not settled about him yet; the man was young, and everyone had their flaws. But he would have to consider him carefully, because it would be Angstrom's reputation on the line for having selected him if he did. When he got back up to his apartment unit, he cleaned up a little before going to bed, and he thought about what a fascinating case Rand was with her sexual dysfunction. But he also told himself that her rehabilitation was not his calling, or something that he even had enough time and energy to address if he wanted to. There was something good inside of her, he could tell; it was buried deep down inside of her, but it was going to have to be someone else who pulled it out of her.

All of these issues—Keplar, Rand, the PrC submission itself—

served as a good distraction for him, because at that moment, the most important aspect in his life was occurring on the other side of the world, in St. Petersburg, Russia, and he was helpless, unable to do anything to further the cause. He had to wait for events to unfold there between Gordon and the Professor...and the Professor's undeclared accomplice.

18

A Professor's Preparations for Final

It could not have worked out better, the Professor thought, short of the man actually killing Raskalnikov. The Professor came to the conclusion, almost immediately, that the man was a consummate professional—two clean, silenced shots, both hitting their marks, was evidence enough. And his identity, even his physical appearance, went undetected by Raskalnikov. The Americans were serious; that was clear and without a doubt.

The one-bedroom apartment that he was hiding in was rented many months ago. He was nearing the end of a six-month lease, but he knew that if things went according to plan, it was only going to be for another week that he would need the place anyway. The only furniture in it was a chair, a bed, and a small table. That, in addition to a few of his favorite books to pass the time, was all he needed.

If there ever was a point of no return, he had certainly crossed it. After that incident in the alley, he knew that he would never be able to go back to his house. Fortunately, he anticipated that such a time would eventually come, and he had long ago gone through the process of emotionally detaching himself from his home and all of his most cherished belongings. He made his peace with it all, and in prayer spoke to the spirit of his dead wife (he was still a believer), seeking courage and affirmation for what he was doing. After all, it was not just *he* that was fleeing to another country — the country that for so long had been his country's central enemy — it was that he was taking another on the journey, and placing that person in a potentially fatal situation.

His thoughts shifted and went back to Dmitri. Their relationship was now definitively severed. He had been blind to Dmitri's corrupted and distorted soul for so long, reluctant to believe that someone he had loved so much could turn against him. But it was all clear to him now.

The signs of Dmitri's deviation were subtle at first, but they were there — little laughs behind the Professor's back, slight challenges publicly of his work, and new allegiances with those who were not aligned with the Professor's views. When Dmitri first started to become distant those several years ago, the Professor could not understand why, and it bothered him and hurt him greatly. Eventually, the Professor's wife told him that it was just the signs of a young man entering manhood and wanting to set his own path. Perhaps, she said, if they were completely honest with themselves, Dmitri did not feel comfortable or even agree with some of the political positions that the Professor was taking, and that was absolutely Dmitri's prerogative, for he was a man of his own convictions, and was not so indebted to them that he should jump off of a cliff just because the Professor chose to do so (though she was not quite so blunt).

It was not until that one day, however, some time after the death of his wife, that the Professor knew that there was a possibility that it was much more than that: for the briefest of moments, the Professor *thought* he saw Dmitri together with Beak-nose, exchanging words with the man, even laughing with him.

Did he really see it, or was it just a dream, imagined during his great period of grief? The supposed sighting happened when the Professor was at the very nadir of his depression over his wife's death, and so he was afraid that his mind might have been playing tricks on him. After further observation of Dmitri, he could still not tell for sure whether there was any kind of an association between the two of them or not, but thoughts of it ripped him apart. He began to wonder how long Dmitri and Raskalnikov might have known each other. How long before his wife's murder?

It wrenched his heart and taxed his mind to no end. The thought of the two talking to each other was like a knife to his heart, intensifying his feelings of hopelessness to a level beyond description. And so, it was with great difficulty when he came to the decision that he must create the contingency plans—plans to account for the real possibility that his protégé had turned against him, and was aligned with him no more.

* * *

It was very shrewd of the Professor to have Dmitri buy the train tickets for them so far in advance. No doubt the ticketing agency would have been alerted by now, after the incident in the alley, to be watchful for anyone attempting to purchase such tickets by a man fitting the Professor's description (or credentials), so the Professor was glad to already have them in his possession. As it was, he had to pay off a certain train conductor handsomely, and well in advance of the inevitable KGB bulletin that would be issued on the Professor, in order to entice the man to permit entry onto the train when the time came.

His plan also called for surveillance to be installed on the train. That would allow for his new American friends to see what transpired on the train real-time, and keep them fully engaged so they would take their new visitors with open arms. His instructions for installation of the surveillance were precise, and he did not rely upon any student to do this particular task for him; it was too important, so he had one of his oldest, most trusted friends do the work—a professor who stood by him through thick and thin, one of the few remaining. The specially developed micro-camera would need to be installed during the last train run of the preceding evening. That way, it would have the least chance of detection, because there would be no more runs between the time of that installation and the first run of the train— the one that the Professor would take the very next day. Fortunately, the train out of St. Petersburg was equipped with WiFi, so it would be an easy effort to have the images transmitted over the internet. He knew that the WiFi access would be restricted, but developing a simple algorithm to get past the firewall was nothing for him. It was a gamble to directly send the video feed from the train's ISP routing, but that was just one of many gambles involved with the final steps of his plan. Up until then there were many redundancies built into his plan to account for things that might go wrong; he knew, however, that that would not be the case anymore. The decision tree had much fewer paths left to it, and fewer opportunities for contingencies to account for unforeseeable deviations.

The whole trip would start with a train from St. Petersburg to Pskov. From there, entry would be gained into Tallinn, Estonia by bus, which would involve a dangerous border crossing. If that transfer was made successfully, the rest of the trip would still be dangerous, but not nearly as much. From Estonia it was on to Latvia, Lithuania, and finally, Poland. He would not be truly safe until they reached Poland. The trip would seem like an eternity; from St. Petersburg to Tallinn alone it would take seven long hours.

From first entry into Poland, an escort was supposed to be there for transit into Warsaw, and from there, safe passage into the United States. That part of the planning was Angstrom's responsibility—the flight plan out of Warsaw.

The Professor figured that the St. Petersburg-Pskov-Warsaw route would be much safer than any attempt to go through Finland. The ride to Pskov, on an early Sunday morning, would be much less traveled and populated than anything to Finland. It would comprise the best chance to minimize the number of officials that might be about, and those officials that were on his selected route would likely not be as well trained and sophisticated as those associated with an attempted defection via Finland.

No conventional firearms could be carried by him onto the train; he could not envision a way to do so without there being a great risk of detection at a security checkpoint. He would not even risk bribery to do so. Therefore, he would have to do something a little more unconventional and devise something special to suit his needs. It was a simple task for such a genius in military technology. It amounted to nothing more, really, than the building of a new toy. The only challenge was to craft it so that it would run off of the charging port built into the seat's armrest on the train without tripping a fuse. A simple voltage doubler inserted into the modified electric shaver, along with a current multiplier, would do the trick; the effort was straightforward enough for him, and even a man as old as him could wield such an instrument ably enough in order to make it dangerous.

He went through how he envisioned it all unfolding countless times in his mind during his stay in the apartment that final week. Over and over again the imagery flowed as to how the event might play out. Finally, he had to find something to distract himself from thinking about it; otherwise, he would never get any sleep. Time had to pass, and a mental escape was necessary in the meantime. Reaching inside the old, green, military bag, he pulled out his copy of *Crime and Punishment*; the one with all of his wife's notes and comments scribbled inside of it. Of everything that was still left in his old home, that was one of the few remaining possessions that he still cherished; one that would be carried with him into the New World — if, in fact, that was his destiny.

When he opened its front cover, he found the old, sepia-colored photograph of his wife that he had placed there. He looked at her intently, and tears came. He closed his eyes, brought the picture close to his mouth and nose, and held it there for a moment, smelling the old musty scent of the paper. Then he opened the book to the first chapter and began reading: *On an exceptionally hot evening early in July a young man came out of the garret in which he lodged in S. Place and walked slowly, as though in hesitation, toward K. Bridge....*

The words worked like aspirin on his brain, and he began to calm down; his nerves settled, and his tension abated. Classical music and literature always had that effect on him. Whatever happened in Russia, and wherever he might be afterwards, he always took comfort in the fact that those great works of the past would always be with him, as his spiritual sanctuary.

* * *

After the incident in the alley, the safe house did not feel so safe anymore. The next one-and-a-half weeks were going to be difficult as Gordon waited for the appointed day in the Professor's plan for their departure. He was going to have to sweat it out in that house, however, and hope that it would not be discovered for what it was. There was no other choice.

KGB forensics would be studying any bullet casings or shrapnel they found from his gun, he was certain of that. It was standard procedure. He knew when he was standing there in that alley, with the incapacitated body lying at his feet, that he should have looked for the casings, and tried to dig out any shrapnel that might have been still lodged in the body, but it would have taken too much time, and would therefore have been too great a risk. He had to get out of there fast.

The scene kept playing itself over in his mind. He still struggled with the fact that the man he shot was not the same person as the one in the picture—the one that the Professor provided to him in stipulating the assassination condition. So why did the Professor lead him to that man in the alley? There was most certainly a purpose in it, but it was not obvious. Maybe there were *two* KGB men hounding the Professor, he postulated.

But then why not ask for both of them to be taken out? The more he thought about it, the more he began to think that maybe the man in the alley was the one that the Professor really wanted to have assassinated, and the other man, the one in the photo, was someone else. A long time nemesis, or rival?

It was cool in the house, even up in the attic where he spent a lot of time, but Gordon did not want to turn the heat up too high, in order to minimize the amount of steam exhausted through the exterior ventilation of the house. If the house was not used regularly, which he was sure was the case judging by its appearance, then steam coming out of it all of a sudden would look peculiar and certainly attract attention.

The same was true for turning on lights in the house. In investigating the minimal number of lamps that were there, he saw that none of them were on timers, so they probably remained off the majority of the time. He was not about to call attention to the place by lighting it up.

It was on one particular occasion, rather late in the evening as Gordon sat in the attic reading under the light bulb, that he heard something that sounded like a knock at the front door. He remained perfectly still and listened, wondering if perhaps it was not a knock at all but just a tree branch or some debris that had been blown up against the house. Then he heard the knocking again.

Instinctively, he reached for his gun, made sure it was loaded, slipped it into his shoulder holster, and put on his jacket. Then he quietly went down to the first floor, wondering who it could be so late at night. For all practical purposes, the house should have looked like nobody was home, or that anyone who *was* home was asleep, with all of the lights turned off and it being pitch black outside.

He crept to the back of the house and ever so carefully pulled the bottom corner of a window curtain away so that he could look out and try to determine if anyone else might be there waiting to ambush him if he tried to make a break for it. He couldn't see anyone, but not wanting to take any chances, he went to a window at the side of the house and quietly unclasped its latch. Slowly he opened it on its hinges, inward toward the inside of the house, and then, after peering out and looking in both directions, maneuvered out of it, landing quietly on the ground outside. He pulled his gun out and crept toward the side of the neighboring house, slowly moved alongside its outside wall toward the front, and gained cover under a clump of bushes. He remained hidden there and tried to see through the bushes who it might be at the front door, and what was happening. An old shopping cart filled with junk was parked on the sidewalk in front of the house, and at his front door stood what appeared to be a homeless person. The man was shabbily dressed with a scraggly, long beard, disheveled hair, and a ragged hat on his head.

Gordon was still not satisfied. He needed to wait and observe further in order to determine whether it was truly a homeless person, or instead, a KGB agent in disguise. He knew it wasn't the Professor; the man was too short and stocky, and the Professor did not know the location of the safe house anyway.

The homeless man moved on the porch to a front window and cupped his hands over it to try and look inside. Fortunately, Gordon had kept all of the heavy curtains drawn so that there was nothing that could be seen. Maybe it was the complete darkness that attracted the man, and he had been scoping it out over a period of time, targeting it as a place to squat. If that was the case, then of all the things that Gordon thought he might have to contend with on his mission, a homeless person looking to camp out at his safe house was the last thing he would have contemplated. Gordon wondered how long the man was going to stand there if no one ever answered the door.

Finally, after enough time had passed with nothing happening, he decided to take matters into his own hands. He was going to have to take a chance, because he couldn't think of anything else short of waiting there and hoping the man would eventually just leave, and he didn't know how long that might take.

He put his gun back into its holster under his jacket and crept toward the back yard, all the while making sure that no one was around to ambush him. When it looked clear, he came out from the bushes at the back and quickly ran around the backside of the neighboring house until he was on the other side of it. From there he went alongside it toward the front and eventually came out onto the front sidewalk. He was able to do all of that unseen, and he began walking on the sidewalk toward his house, with the appearance of being just an average pedestrian out for a walk. If the homeless person had been casing that area for a while, then there was a chance that Gordon would stand out in the neighborhood; he was risking it.

When he reached the shopping cart, he stopped to look at it. Then he looked up to his porch where the homeless person was and saw that the man had not noticed him yet.

"Pssst," he said. "Pssst."

The homeless man heard him and turned around.

Pointing at the shopping cart, with his finger moving up and down while he did so, Gordon said in a half-whispered voice, but loud enough for the man to hear, "It looks like you've got some good stuff in here." As he spoke in Russian, he utilized all of his linguistic training to carefully mask, as best he could, his American accent.

His ploy seemed to have its intended effect, for the man began to come down from the porch over to him. Gordon stood ready in case the man was someone other than who he appeared to be, and watched carefully as the man approached.

"You want to buy something?" the man asked in Russian.

"No, no. I was just admiring it. What brings you around here?" Gordon responded.

"Got lost. You live around here?" the short man asked.

Gordon was not about to answer that, and he didn't want to stand out there in the open for very much longer either, so he acted quickly. He reached into his pocket and pulled out a few Rubles.

"Here," he said as he offered them to the man. "I think there's a small store around the corner, a couple of blocks down. Get yourself a little something." He pointed in the direction where he had just come from, so that if the man followed the suggestion and walked in that direction, Gordon could keep walking forward in the opposite direction , eventually turn into a neighboring yard, and circle back to his house.

The man did not hesitate to take the money. He was unsure as to what to say next, until finally he said, while pointing to the things in his cart, "You sure you don't want anything?"

"Yeah, yeah, I'm sure. You keep it. Have a good evening." He patted the man on the shoulder and began to walk forward, signaling to the man that their meeting was over. Gordon could hear the wheels of the shopping cart start to move along the surface of the sidewalk, and after another step he quickly turned over his shoulder to ensure that the man was walking away in the opposite direction.

The situation was resolved, as best Gordon could tell. There was still a chance that it was all part of an elaborate trap—that the man was just scoping him out—so he didn't immediately return to the house. It was chilly outside, even with his jacket; the chill hit him right away, now that the homeless man no longer occupied his attention. He made his way back to the bushes where he was hiding before, and he stayed there for another hour, motionless, observing the area to make sure that no one else came around that looked suspicious.

It was a little bit of a reach to grab the sill of the opened window, but he was able to get to it with a slight jump, and with a small effort he pulled himself back into the house. The wait outside in the cold, along with the stress of the situation, caused it to take some time before his nerves settled. He continued reading his newest book, *The Miernik Dossier*, for another couple of hours to relax himself. The book was a good distraction and helped calm him down. Eventually he turned off the small attic light and went to bed.

19

She Isn't What She Seems

Vitebsk Station, a historic building located in the southern part of St. Petersburg, was where the Professor would go for the train that ran to the Baltic States. Obtaining his fake passport was one of the easiest facets of his plan, thanks to his having Dmitri obtain it for him a long time ago. Security was light and amateurish, just as he had predicted, and his passage through security was a non-event. Everyone had been paid off as necessary, beyond the knowledge of even Dmitri. The old military bag he carried made it through the x-ray scanner without a hitch, and he passed through the full body scan without objection. The agent at the x-ray monitor did not see anything unusual about the modified electric shaver in the bag.

It almost seemed too easy for him; he thought Dmitri's alerting Beak-nose about his itinerary would have caused at least a few more armed guards than what he saw. Either way, it didn't matter; Dmitri may have known the station from which the Professor would depart, and even the train he would take, but he didn't know his manner of entry, or who he paid off for it.

The Professor could have changed his travel plans after Dmitri's betrayal was confirmed—a different station, different train, even a different route out of Russia—but it would have been difficult to get another set of tickets at such a late stage, especially without Dmitri's help. He also could have made alternate travel arrangements well in advance, as one of his contingencies. None of those options figured into his overall plan. He *wanted* Dmitri to know, so that Beak-nose would know. That was absolutely necessary if the Professor was going to have any real chance of achieving all of his objectives. Therefore, the main portion of his escape route remained unchanged—at least the first part of it.

He just wondered if the hatred in Beak-nose, his narrow-minded nature, and his fixation with wanting to be the one to stop the Professor, would be enough to cause him to be there himself; to insist that it be so, even despite his having the new, physical handicaps he now had. The Professor predicted that it would. His next consideration, then, was whether Beak-nose's superiors would permit it. For all the Professor knew, they could have decommissioned the man after the embarrassing incident that occurred in the alley. Then again, perhaps Beak-nose did not even inform his superiors about this part of the plan, or that today was the appointed day, in which case, Beak-nose would just show up, without his superiors even being aware of it. The Professor realized that he was letting himself get caught up in his decision tree analysis, and he forced himself to stop; he couldn't afford to be distracted.

Without trying to appear obvious, he looked around periodically to see if he might be under observation. He purposefully showed up as close to actual departure time as possible to minimize the time he would have to spend out in the open before boarding began, especially when people suspected that he would be there. He needed to get on the train; that was crucial. With the bribes he paid, he was able to bypass the gate area and wait for boarding alone in a small, nearby office. It was still a nerve-wracking exercise in patience, even for a man of his advanced age. There was no sign of Dmitri so far, or Beak-nose, and he kept watching for them through the slats of a screen that was pulled down over the window in the office that overlooked the general boarding area.

* * *

As early in the morning as Gordon needed to be awake on that day, he was up even earlier. He had gotten more than enough sleep, and he was not about to take any chances by not getting up as early as he did. It was so early that it was still pitch black outside, and a street lamp at the corner of the block was the only illumination in the immediate area of the safe house. When he peeked through the curtains to look out of the front window, no one was in sight. The same was true when he peeled away the curtain to look out of the back of the house. Still, he wanted to take every precaution, and even after everything looked clear at both ends of the house, he went out of the same window as he did before, from the side of the house.

He started to walk through the neighborhood toward the business district so that he could catch a taxi to get to his first destination, but before he walked too far, he turned back around to take a final look at the house. It would likely be decommissioned — the last time that anyone from his organization would ever use it.

When the cab stopped a few blocks from where he needed to go, he stepped out into the chilly air and looked around. There was no traffic, save for a few taxis here and there making their rounds, and there were no pedestrians about. It was desolate, almost like a ghost town. No signs of life. His walk was brisk as he kept both hands in his pockets and his head down, trying to look as nondescript as possible.

He had already surveyed the whole area in advance over the last few days, and as a result, he knew every nook and cranny where someone could be hiding, waiting to lurch out and overtake him. When he neared the old high rise building, he momentarily remained at a distance to look through the ground floor windows and scope out the front lobby. There was no movement. He looked around the general area outside again and could see nothing suspicious—no car parked with someone waiting inside of it, no one lurking around the front entrance. He stayed hidden near the base of the old building for a while longer and continued looking in all directions to make absolutely sure that no one was there staking out the area.

The badge that the Professor had provided him, from the same metal box he found in the forest, caused the electronic lock of the glass front doors to unlatch, and he entered an empty lobby, save for a security guard sitting at the front desk. The man had his head down on the counter, sleeping, while a small black-and-white television near him played. Gordon walked right past him to the elevators, and when it stopped on the third floor, his hand reached to hold the door open as he cautiously peered out in both directions to make sure it was safe.

Dmitri's apartment was number 308, the fourth door on the right. Before trying to enter it, he reached for the ceiling light close to the door and quietly broke its bulb (it was too high to unscrew), thereby darkening the hall so that a bright light would not shine into Dmitri's apartment when Gordon opened the door.

Gordon had already verified a few days ago that the key the Professor had given him still worked on Dmitri's door, and he was counting on it working again. The last thing he wanted was to have to pick the lock of what he thought was a KGB agent's apartment door at three o'clock in the morning. First he put his ear to the door to make sure he could not hear any activity occurring inside, and then he pulled the door further against its frame while it was still locked so that when he turned the key it would make the least amount of noise. The Professor assured him that there would be no alarm system, and Gordon was surprised to discover that it was true during his earlier reconnaissance of the place.

There was only the minutest clicking sound when Gordon

unlocked the door, and he slowly began to push it open. It was completely dark inside, and after Gordon went in and shut the door, he paused momentarily to let his eyes adjust to the darkness. He remained crouched in the front entryway for a while to wait and listen for any sounds, and then he pulled the silenced M9 out of his jacket and made his way toward the bedroom. There were still no signs of life as he went through the main living area. Everything was as he remembered it from his earlier visit, and by now his eyes had adjusted enough so that it was no trouble at all to make his way to the bedroom in the darkness.

He pointed his gun, crouched low, and carefully entered the bedroom. He thought he could make out a figure lying in the bed and walked over to the side of it. The silencer on the barrel of the M9 was almost touching Dmitri's temple when Gordon reached over to the bedside lamp with his other hand and turned it on. Gordon's eyes adjusted quickly, but for Dmitri, his eyes reflexively clenched tighter in the sudden light. When Dmitri finally realized that the lamp had been turned on, he opened his eyes and squinted to see what was happening.

Sensing the gun against his temple, a feeling of terror came upon him, and he instinctively tried to sit up. Gordon used his free hand to push down on Dmitri's forehead, and he pressed the barrel of the gun harder against his head as he said, in Russian, "Stay down."

There they remained, in silence, looking at each other, with Gordon towering over him. He studied Dmitri's face, confirming that he was indeed the same man as the one in the picture that the Professor had provided.

"What do you want? Anything in my apartment is yours...you can have it... please, just don't hurt me," Dmitri cried in Russian.

Gordon said nothing.

Reality was setting in for Dmitri; he was getting his bearings. "Did Raskalnikov send you here?"

Gordon shook his head slightly to indicate no.

Dmitri thought further, and as if a light bulb went off in his head, his eyebrows raised and his eyes opened wider. "The Professor?"

This time Gordon nodded yes.

"Who are you?"

No response. One should talk as little as possible in such situations. But then Gordon relented and said in Russian, "Don't you know?"

"British?"

A slight, sarcastic grin appeared on Gordon's face, but he was not about to clarify the man's confusion any further.

"Turn on your stomach, put your face into the pillow, and both of your hands behind your back."

"Please, don't hurt me," Dmitri pleaded.

"Just turn over," urged Gordon, gritting his teeth to make it clear that he was serious.

As Dmitri did as he was told, Gordon reached into his jacket with his free hand, pulled out a pair of plastic, cable-tie handcuffs, and bound Dmitri's hands.

"Get out of bed and lie on the floor, face down."

Dmitri could not believe what was happening. It was as he feared after he learned what had happened to Raskalnikov. The Professor somehow learned of Dmitri's allegiance to Raskalnikov and had cut him out of his plans. But he never would have suspected that the Professor would resort to this.

As he got down on the floor, the terror he felt intensified; he feared for his life. Then he recalled Raskalnikov's description of what happened in the alley, and how someone had shot him with such precision. The recollection suddenly caused Dmitri to panic. He feared that this was the same hit-man and that he was going to be tortured. He wanted to scream in fear.

Almost as though he could read Dmitri's thoughts, Gordon said, "Don't make a sound or I'll kill you, instantly." Then he holstered his gun and bound Dmitri's ankles with another set of cuffs.

Now would be the hard part, Gordon thought. Hard only because the man was so tall, slightly heavy set, and appeared to be as stiff as a board—the body would not be used to the bending that it was about to experience. Gordon put his knee into Dmitri's back, grabbed the tie around Dmitri's hands, and pulled it backward until the bound hands were close enough to the bound ankles so that he could attach them together with a third set of cuffs. Dmitri was hog-tied.

Now that Dmitri would not be able to move, Gordon took his gun back out. Dmitri rolled onto his side to relieve the stress on his limbs and body as best he could.

Gordon kneeled down on the floor, leaned close to Dmitri's face, and pointed his gun at it. "Who are you?"

It was the last question in the world that Dmitri expected. How could this man break into his apartment and do what he was doing without even knowing who he was? What had the Professor told this man?

"Do you want money? I can give you everything that I have?"

"I said, 'Who are you?' " Gordon pointed the end of the silenced pistol into one of Dmitri's eyes, and the eye closed reflexively. Gordon then pressed the gun on the closed eyelid and applied pressure.

"Dmitri Pavlovitch."

"KGB?"

A perplexed look appeared on Dmitri's face. "No. I'm not KGB! I'm a University Professor!" he said in a desperate tone. "You...you have the wrong man!"

Gordon was surprised at the answer, but he maintained his composure despite the unexpected response. Thoughts of a double-cross, or that something was not right, momentarily went through his head, but he ignored them. It was too late to change course now, and he had his orders. He did what he was trained to do in such situations: when he was not sure of what was going on, he would ask open-ended questions to try and extract more information from his captive.

"Why?" He pressed the barrel into Dmitri's closed eye even harder to stress the point.

Dmitri was confused. "I..." He was not sure what he was

being asked, and therefore what he should say. "I love the Professor...but, he's at the end of his life. He's pressed things too far. I don't think...the Professor just never stopped...and now it's too late for him."

It was gibberish, but there was enough there to make Gordon feel better about what he was doing, and that he had the right man. He was never going to kill him, not even when he thought he was KGB. Learning now that he was not even KGB, but only a civilian, reinforced his organization's decision. There must have been some reason, though, that the Professor made the man out to be KGB, or at least, requested his assassination. Maybe he never wanted the man assassinated in the first place. Maybe he just wanted him...detained.

Gordon looked into the one opened eye of the man for a moment, and all he saw was fear. With one powerful blow to the head with the butt of his gun, he knocked Dmitri unconscious. Before he left the apartment, he put some specialized tape over Dmitri's mouth—a precisely sized strip he had brought with him—and cut all of the phone lines.

Now it was on to his next destination: Vitebsk Station.

* * *

The Professor would not feel the least bit comfortable until the train started to move out of the station. His heart raced as he sat in one of the train's personal compartments and waited to see if Dmitri would arrive, or anyone else. The compartment had two bench seats, and he had instructed Dmitri to reserve the whole compartment so that he and the Professor could theoretically be in it together with the rest of The Mighty Five. The Professor slid the opaque panels closed over the windows for more privacy.

There was nothing that the Professor had brought with him save for a few clothes in his bag, along with the modified shaver, and his cherished copies of *Crime and Punishment* and *The Brothers*

Karamazov. There was nothing valuable about the books themselves, except for the fact that they were the copies, in Russian, that he and his wife had both read and cherished between them. He slowly flipped through the pages of *Karamazov* in an effort to calm himself. Her personal, handwritten notes were interspersed throughout it. The Professor and his wife had had many long, deep discussions about that book while sitting in front of their fireplace together on cold, winter days. He remembered how she had told him that the first time she read the tale of *The Grand Inquisitor* she had to stop, and could not pick the book back up for days, so moved was she by its power.

He was jarred from his thoughts when the train began to move and pull out of the station. Dmitri had still not arrived, and the Professor experienced a pang in his heart, knowing that he was leaving the man forever. He wondered if any pain had been inflicted upon him while he was detained.

Gordon was able to board the train in time, and he walked down the aisle of another car that was two down from the Professor's. Most of the seats were empty; four other passengers were in Gordon's car, whereas, at full capacity the car was able to seat thirty passengers. Unlike the Professor's car, which was comprised of only four private compartments, each with its own door, Gordon's car had only a common seating area. There were two columns of cushioned bench seats, each one with room for two people. Gordon sat in one of the middle rows and put his ticket under the metal tab that was on top of the backrest of the seat in front of him.

The Professor heard a noise at the door of his compartment, and he looked up to see who it was as the person entered. It was the train conductor, who had come to check his ticket.

"Good morning, sir."

"Good morning," the Professor replied as he looked up at the man with a slight smile.

"It looks like the other patrons of your compartment decided to not show up," the conductor said as he looked around the empty compartment to emphasize the point. The door had closed behind him after he entered.

"Hmh? Ah, yes, it seems so," the Professor said as he reached inside of his coat for his ticket.

As the conductor began to punch holes through it, he said, "Odd ride this morning."

"I'm sorry, I don't know what you mean."

"This whole car is sold out. Not just this compartment, but the whole car. That's pretty unusual for the train this early on a Sunday morning—all four compartments. The whole train is maybe at thirty-percent of capacity, but this car is completely sold out."

The Professor just nodded, not sure what to say.

"What's even odder, though, is that other than yourself, none of the other passengers for the car bothered to show up."

Now a look of concern appeared on the Professor's face. "Really?"

"Yeah. Yours is the last compartment in the car, and all the others are empty. I just came from them. Pretty strange."

"Yes... I agree."

The ticket checked, the conductor looked down at the Professor for a moment longer and said, "One more thing. My instructions are to not disturb the people in this car after I punched their tickets. I guess they were supposed to be very private, important people...or something. Anyway, although no one else showed up, I'm still going to follow instructions and not come by here anymore. But, if you decide for some reason you need me, don't hesitate to press the call button. Alright, have a nice trip, sir."

"Thank you. Thank you very much."

When the conductor left, the Professor reached into his small bag that was resting on the seat beside him and pulled out the shaver. It was no accident, he thought, that the car was sold out, and yet no one was there. He plugged the special cord into the shaver, and the other end of it into the power jack built into the armrest. Then he put the shaver down on the armrest and kept his hand over it. If anything happened, it would be his only means of defense. He looked up at the corners of the compartment ceiling to look for the micro-camera, and he spotted it; his old friend did not fail him. He still felt a sense of urgency and alarm because of what the conductor had told him: the entire car was empty. He was there all alone, vulnerable, and the conductor would not be coming back, despite how long the train ride was going to be.

Raskalnikov and his two Angels of Death were in the train car adjoining the Professor's. Other than an old lady that sat in the front row of that car, they were the only ones in it. In terms of order of the train cars, it was the Professor's car, which was the very last one of the train, then Raskalnikov's, and then Gordon's. The train was a conventional model and moved at a moderate speed. There was plenty of time for Raskalnikov to do what he wanted to do, and he and his men did not make their move for the first three and a half hours of the trip. Village after village could be seen through the panoramic view of the windows as the train sped along the track.

According to the Professor's instructions, Gordon was to make his way to the Professor's compartment only *after* the train had stopped at Novgorod and then continued on its journey. The train would be arriving there soon.

Raskalnikov, therefore, would be making his move first, because according to the instructions that Dmitri had given to him, Raskalnikov was supposed to go to the Professor's compartment fifteen Kilometers *before* the Novgorod stop. Raskalnikov tapped one of his henchmen on the shoulder and motioned his head in the direction of the Professor's car. When the man got up, Raskalnikov followed him, still walking with a crutch, and the cast on his right arm resting in a sling. The second henchman trailed both of them. The man in front pulled the latch that unlocked the door connecting their car with the Professor's, and he slid the door open and held it so that he and Raskalnikov could walk through it. The second henchman reached over Raskalnikov's shoulder to hold the door open for himself as they passed through. When all three were in the Professor's car, the door shut, and the second henchman remained standing there in front of the door's window so that no one could see through it.

Raskalnikov motioned to the first henchman to take his post, which was right outside of the Professor's compartment. The other henchman was to remain where he was to make sure that no one could enter the car. Raskalnikov leaned his crutch against the wall, reached into his jacket with his left hand to pull out his gun, transferred it to his right hand, and then grabbed his crutch again. He hobbled over toward the Professor's compartment. The pain from his injured ankle was still great, and he grimaced. The two henchmen wondered why Raskalnikov did not let one of them finish the job for him, but they said nothing.

Raskalnikov stood in front of the door for a moment, recognizing the significance of what was about to happen, and then he cumbersomely grabbed the knob with his left hand and opened the door. The Professor looked up at him immediately, and his grip on the shaver tightened.

* * *

Angstrom was alone in his apartment; it was close to ten o'clock in the evening in the United States, and he had just logged into the internet site that the Professor had previously identified to him. He had no idea what to expect, and when the site came up, it was completely black, except for the following words, in white lettering, that were displayed across the middle of the screen: "Transmission Pending."

At about eleven o'clock, he refreshed the screen, and just then a small window appeared. The PC's media player was buffering data. When it was done, a black-and-white image appeared; it was a video transmission from the micro-camera in the Professor's compartment. Angstrom leaned closer to the monitor to better discern the image and realized that it was the Professor he was seeing; it was a live video stream of the Professor in his train compartment; he was streaming his defection!

It felt eerie for Angstrom to be watching. He was so many thousands of kilometers away, unable to affect anything that happened, not even able to communicate with the man, yet he was watching him on his PC. He continued to study the video stream carefully and considered every detail that he saw. Angstrom noticed how old and frail the Professor looked.

Then it suddenly occurred to him that the Professor was alone. There were supposed to be two people that were defecting. Where was the other person? In another compartment? Angstrom became concerned that something may have happened to Gordon, and that was why only the Professor was there. It was going to be a very stressful night, he said to himself.

He saw when the train conductor came into the compartment, talked to the Professor, and punched his ticket. After they exchanged words and the man left, he saw the Professor take out something and put it on the seat's armrest, but Angstrom could not make out what it was. The image resolution of the feed was not high enough. Then he saw the Professor plug the object into a power outlet, and he was perplexed. How odd it seemed to him in that he thought he recognized the object as an electric shaver. Surely the Professor was not going to shave. He could only imagine the absolute mess that would result if the Professor decided to shave right then and there.

Nothing happened for quite a while. All that Angstrom saw was the Professor sitting alone and reading from an old book, electric shaver resting next to him.

Eventually Angstrom saw the door to the compartment open and another man enter. He thought it was a strange sight: the new man had a cast on his arm, and another on his left leg. Angstrom looked carefully and saw a gun in the man's hand, and it was pointed at the Professor!

* * *

As Gordon sat in his seat, he continued to look out of his window. The old villages he saw pass by looked serene. More importantly, he watched the signs posted at each of the villages as the train stopped at them. He needed to be absolutely sure when it was time to make his move.

When the train began moving again after stopping at a local station, a woman walked through his car. She came straight down the aisle and kept walking right past him. She looked eerily familiar to him, and he realized in an instant that it was the same woman that had suddenly sat down at his table that evening in the restaurant. He was sure of it! He became uneasy, thinking the situation incredibly odd, and that there was no chance of it just being a coincidence.

With a determined gait, she passed through his car and into the next—the one where Raskalnikov and his two henchmen had been at the beginning of the trip. That car was completely empty by then, and she walked down the center aisle of it until she reached the door that led to the Professor's car. She slid the door open, and the first henchman immediately spun in front of her and said, "This car is closed, ma'am. You'll have to find another." He looked at her with a very businesslike expression, and then he pulled the door closed.

It was her good fortune that she was trained to be left-handed. She reached into the large purse that hung over her right shoulder, and with her left hand grabbed the military knife that was in it. After that she tossed the purse onto one of the empty seats so that she would be unencumbered. Now was the critical moment. With her right hand she ever so slightly slid the door sideways to open it a crack. She leaned forward to poke her face into the opening, with her left hand remaining to the side behind the wall, so that the knife was hidden from view.

"Psst, excuse me."

The man turned to her again, and in order to hear what she wanted to say, he leaned his head close to the cracked door to meet her gaze. As he did so, she quickly pulled the door open further, and he never saw the blade that swung up from underneath and went straight through the carotid artery of his neck. He made only the slightest gurgling noise before blood spewed forth from his neck, splattering onto the woman's face and chest as she pulled the blade back out. He fell straight to the floor with blood continuing to spew out.

She saw that there was another man at the other end of the car. He had seen his partner drop to the ground with blood squirting from his neck, and as he turned to walk toward his partner, he reached into his holster to pull out his gun. Almost simultaneously, with a constant, fluid motion, the woman pulled the door all the way open, stepped over the body lying on the floor, and entered the car. The man that was coming toward her was startled at the sight of a beautiful woman covered in blood, and the distraction was just enough for him to not notice the knife in her left hand. She cocked her arm back with precision, in accordance with all of her exhaustive training, stepped forward with her right foot, and flung the knife directly into the man's chest, piercing his heart. It was a throw of approximately twenty feet, and a direct hit.

The pain shocked him, as well as the surprise at what had just happened. He was not sure what to do. Should he leave the knife in his chest or pull it out? He knew he was going to die—the pain was excruciating—but within a matter of microseconds, he struggled with the question: What would give him the most time, so that he could kill her? With one hand gripping the knife's handle, he raised his other to point the gun and shoot her.

When she saw his gun rise, she crouched down low, back toward the first man that was on the floor. There was not a lot of room for her to move, and she was too far from the door to try to go back out of it before a shot was fired.

The hand holding the gun waivered, and the first shot flew over her shoulder. She tried to turn the body over that was lying on the ground in order to get to the dead man's gun, but it was too heavy, and there was not enough room to maneuver. The other man by that time had fallen to his knees, struggling to remain conscious, and was preparing to take another shot. She knew he was going to get it off, so she did the best she could to time a jump to the other side of the hallway; there was not much room because the private compartments took up so much space in the car. The man's gun let out another "thwirp" from the silencer, and a bullet grazed her thigh.

After that, the man fell to the ground, unable to maintain consciousness any longer. His body fell directly onto the knife lodged in his chest, thereby driving it further into his body. Before she did anything else, she dragged the body of the first man all the way into the car, so that the door separating the two cars would close.

Miraculously, between the silent deaths of the two henchmen delivered by her knife, and the suppressed noise from the firing of the second man's gun, virtually no noticeable noise was made during the whole series of killings; the Professor and Raskalnikov heard nothing, and the door to the Professor's compartment remained closed. She looked down the aisle for a moment and steadied her nerves as best she could. She was so close now.

At that same moment, Gordon could not get over the fact that he saw that woman walk past him on the train. He knew that time was growing short, and that soon the train would be in the Novgorod station, but he was determined to get up and go to the next car to find that woman. He cautiously walked over to it, full well knowing that he was breaking with instructions. When he made it to the next car, he stood at the end of it and looked around, but it was empty. He slowly walked down the center aisle, well aware that someone could be lurking within one of the rows of seats to try and jump out at him as he went past, so he remained alert as he made his way toward the last car on the train—the Professor's car.

Meanwhile, that woman, now all covered in blood, was standing outside of the door to the Professor's compartment. Having checked the other three compartments and finding them empty, she knew he had to be in that last compartment. She had the gun that she had taken from the first dead man in her left hand now, and she flexed her right hand several times to relax its muscles before she turned the knob to open the door.

Angstrom was watching the video stream; he saw the Professor, still sitting in his seat with his hand on the electric shaver, and the strange looking man sitting opposite him. Angstrom clearly saw that the two were talking, but he could not hear anything because there was no audio accompanying the stream. He could barely make out the other man's face because of the graininess of the feed. The most distinguishing feature about him, besides the oddity of the two casts and the gun in his hand, was the nose that protruded from his face. It reminded Angstrom of the beak of a hawk.

Then Angstrom saw the man stand up and go toward the Professor. He hovered over him and brought the barrel of the silenced gun to the side of the Professor's face, right up against his cheek. Angstrom wished there was something he could do, and he wondered where Gordon was. He needed radio contact with him.

As Raskalnikov held the gun to the Professor's face, he heard the door to the compartment open, and he quickly became annoyed, thinking that it was one of his henchmen coming in to bother him.

But it was the woman.

"You!" he said in amazement.

At that very instant, the Professor hit the button on his shaver that initiated the high voltage electric charge, and swung it into Raskalnikov's chest. Two things happened next which the Professor had not counted on. First, the sudden shock from the shaver caused a spasm in Raskalnikov's hand that was holding the gun, thereby causing him to involuntarily pull the trigger and shoot the Professor in the side of the face. Blood and flesh splattered onto the compartment's wall, and partially onto the woman. Second, the high voltage that the Professor applied to Raskalnikov's chest was directly over his heart, causing it to go into arrest; Raskalnikov had a heart attack.

"Papa!" the woman shouted.

He was slumped over, with blood flowing out of his deformed, lacerated face.   Her attention was momentarily drawn to Raskalnikov, who was lying on the floor and clutching his chest in pain. She pointed her gun directly into his forehead, and with his eyes looking directly into hers, she shot him. Blood shot upward at her, and now she had the blood of three men splattered on her. She was drenched in blood. She dropped the gun and went over to sit next to her father, cupping his head within her arms.

His eyes turned from the now dead Raskalnikov over to his daughter. "I...love...you," he said with great struggle, looking up into her eyes. "Be...free."

He reached up slowly and gave her the bloodied book that he was still clutching, and with that, died in her arms. She put her cheek to his head and held it close, squeezing it as she sobbed.

Angstrom, after seeing what had just transpired, was stunned. He had witnessed the whole event, but was helpless and could do nothing, and he continued to watch as the woman stood up and gently slid the Professor's head down onto the seat in a laying position. Then she reached over, grabbed the bag that had been resting on the seat next to him, and put the book inside of it. She looked down at her father, and at that moment she was in perfect view of the micro-camera. Angstrom could not believe what he saw: it was her—Candy Mav.

Just then, Gordon swung the door open to the compartment and immediately saw the bloodbath.

"Jesus!" was all he could say at the gruesomeness of the scene.

Angstrom saw Gordon enter the compartment, and after Gordon had a moment to collect himself, he looked up at the micro-camera he knew would be there and shook his head in disbelief.

Angstrom saw Gordon turn to say something to the woman, and they began moving the bodies around. They were going to clean the place up somehow, was all Angstrom could figure.

"All of the other compartments in this car are empty," she said hurriedly to Gordon (in Russian). "Let's move them all into one of them."

"What's the use?" he responded. "If the train conductor comes into this car, he's still going to see all of the blood."

She knew he was right.

"Alright. Let's just move the two dead men that are outside into this compartment. We'll move my father into one of the empty ones; I don't want his body in here with these carcasses. I'll stay in the compartment with my father for the rest of the trip because there's no way I'll get all of this blood off of me. You sit on the seat outside of the door separating this car from the next and don't let anyone enter. Think of something…anything…just don't let anyone in here. We'll be arriving at Pskov soon anyway, and that's where we'll get off."

Gordon shook his head in silent acknowledgement.

20

Transition

Angstrom was disappointed that he did not think to record the video stream, but he didn't know that it was going to be streaming like it did. If he had, he could have reviewed it, and analyzed it; verified what he saw. Now he was going to have to wait for their arrival. It was the middle of the night in the United States, and he was lying in his bed, wide awake.

There was no way he could have been wrong about what he saw; he was trained too well to study and remember faces and details; it was her. The woman he saw on the train was Candy Mav. If there would have been sound transmitted with the stream, Gordon could have said something to him. Angstrom was going to have to wait at least until they were in Lithuania; then it might be possible for them to make radio contact. But more likely, he figured, it would not be until Warsaw that he would hear anything from Gordon.

The killing on the train certainly qualified as an international incident, and Angstrom wondered how it could have happened. Gordon was supposed to have detained the KGB agent, so he wondered who the man was that shot the Professor in that compartment. Angstrom recalled how the man was pretty banged up and could barely walk, much less hold a gun. He speculated that maybe it was the handiwork of Gordon.

He wondered how his organization would take the news that for all the trouble of destroying Gordon's deep cover, and the deaths of those on the train, that they would not receive their expected package, the Professor. *Package*: what a way to refer to another human being, he thought. He figured that the prize they received from the PrC submission would still be deemed compensation enough, but the Professor would have been a veritable gold mine; a person who likely would be able to chronicle decades of Russian military technology; maybe even down to the level of frequencies, critical algorithms, and encryption codes.

Left with nothing but time as he waited to hear from the two travelers, his thoughts turned to her alone. The memory of her image from the video stream played itself over and over again in his mind. She was real flesh and blood. He was trained to not let such players in missions get to him that way, but for some reason it was different with her. Now he knew she was real.

"Christ," he thought, "I haven't even met her."

What was it about her that fascinated him the way that she did? For sure a big part of it was her physical beauty; her physique from the Candy Mav images was exquisite, and nothing about the transmitted video stream dispelled that notion. Her face had a unique look that evoked a sense of distance and coldness, but at the same time it was also striking in its magnificence. He could not wait to meet her and hear her speak. Surely, he thought, her voice would match the beauty and perfection of the rest of her.

He had been involved with beautiful woman before while on missions, so a physical infatuation could not be the only reason for his being enchanted by her as he was. No doubt another part of it was the mystery surrounding her. If absence makes the heart grow fonder, mystery captures it in the first place. There were all kinds of questions he would need to ask her: the overall role she played in the defection, her ties with the Professor, why she chose to help him, and so forth. It intrigued him immensely how a woman of such beauty could have become involved in such a strange and gruesome series of events.

She had a gun on the train; she used it on the crippled man in the Professor's compartment, and it looked like she was covered in blood already when she entered it. There was no hesitation when she held the silenced weapon against the man's forehead and pulled the trigger. Was she a professional?

His powers of observation allowed him to note something else that was significant, even in that short, grainy stream with no sound: she held the Professor in her lap momentarily with compassion, and cried. There was a bond between them, and he guessed that it was more than just a professional relationship. Perhaps they were related.

Questions. So many questions.

He needed to get back to the secure communications room and await word on Gordon's status. There was still a chance that they would not make it across the border, because there were at least four dead bodies on that train that Angstrom knew of (he saw the bodies of the two henchmen get dragged into the compartment), and a woman splattered in blood.

But if anyone could find a way to get them out of there, it would be Gordon. Angstrom wished that he was there.

* * *

Gordon gave her his jacket. That went a long way in covering up her blood-soaked clothes. He wet his handkerchief in the small toilette on the train and brought it to her to wash her face, and on his way back through the cars from the toilette, he convinced a young woman wearing a stylish hat to sell it to him under the pretext that he absolutely had to have it for his girlfriend. She was happy to oblige, especially in light of the large number of Rubles he offered her.

Unless one looked closely, no blood was visible on the woman after she wore the hat with her hair tucked up inside of it, and the jacket that was buttoned up to the top. He sat guard on the seat right by the sliding door, and because that car was still completely empty except for himself, he was tempted to go back into the end car and start asking the woman some questions. Leaving the door unguarded, though, would be too dangerous.

The transfer to the bus that crossed into Tallinn would be the last major risk. After that, it was a matter of tight protection around her at all times until they got onto the flight out of Warsaw. Making such a transition, even with the package that Gordon was escorting, was not normally a notable event for someone as skilled as him, but he knew that there was a high probability that the men on the train were supposed to have checked in with their superiors at some point, and when that did not happen, train personnel would be alerted.

They walked off of the train together, and he wished that he had a woman's coat for her. As it was she was wearing his, which on the train looked fine, but not so much when they stepped off of it; and now he had no jacket for himself, which he felt looked odd given the cool weather outside.

They walked together, his arm around hers, as they made their short walk to immigration, which ended up just being a small area with a few walk-up stations. A dog was there, as was a metal detector, which made Gordon fear that it was not going to be long before the crew on board the train found the bodies, which meant they needed to move fast (he of course had no way of knowing what the train conductor had told the Professor about how no one would be coming to that car for the rest of the trip).

They separated at immigration, and each indicated to the station agent that he or she was traveling alone. Their passports were sufficient for the occasion, and they reconnected some short distance beyond the post. She must have had a damn good passport, he figured.

They still had not said much to each other as they quickly walked away; they needed to create as much distance between themselves and the train as possible. Only one bus a day went directly from Pskov to Tallinn, and it was going to depart just a few minutes from then. They had to hurry and knew that it was going to be tight, but they had to catch that bus.

When they arrived at the depot, the bus was just closing its door to depart, and they moved swiftly to get to it just in time to tap on the door and get inside. The timing was perfect; there would be no need to linger at the bus station, which was good, because it was tiny, and anyone looking for them would have spotted them right away.

The ride into Tallinn was nerve-wracking; there were only a few other passengers on the bus, and the driver tried talking to Gordon and the woman on more than one occasion. They did not respond, however, and the driver eventually got the hint that they were in no mood to converse. Gordon was nervous that the bus would be alerted and ordered to return to the station. Fortunately, that leg of their trip was not disclosed to Dmitri and Raskalnikov, and the Professor had actually designated a different route and destination to Dmitri, so it would not be immediately obvious that they were on the bus, as long as they were not spotted boarding it. In the plan that the Professor had disclosed to Dmitri, they were not even supposed to be getting off of the train at that stop.

When they caught their breath and there was some distance between them and the station, Gordon leaned over to her and quietly asked in Russian, "What's your name?"

She turned to look at him and said, "Anna. Anna Czolski."

Gordon immediately recognized the fact that she had the same last name as the Professor, but he did not comment any further about it.

"It's a pleasure to meet you, Anna," he said, holding out his hand to shake hers. Gordon saw, in his peripheral vision, the bus driver's eyes looking at them in the large, rear-view mirror. There were several rows of tall, empty seats between the bus driver and them, however, so it was difficult for the driver to overhear the conversation or see much of them.

The Professor's plan called for them to catch another bus at Tallinn and take that into Riga, but that was an almost five hour trip, and Gordon and Angstrom knew such a long ride was too great a risk. In actuality, the Professor also knew that such a route was too risky, and he fully counted on his new friends recognizing that fact and making alternate plans from that point onward. When they reached Tallinn, a large, black vehicle with heavily tinted windows was waiting for them at the bus depot. Gordon walked over to the driver's side window and spoke in code to confirm his identity, and after that he and Anna entered the vehicle's back seat.

After the car pulled away, Anna turned to him and said, in Russian, "So, you are an excellent American spy, but we have to stop meeting under such lethal circumstances, hmm?"

He was exhausted, but she made him laugh anyway, which was her intention—even after all she had just been through—and he appreciated the gesture.

The ride would be long, and they allowed themselves to fall asleep. Not much else was said between them for the rest of the trip.

* * *

"I couldn't believe it. I saw everything on the video stream," Angstrom said.

"Yeah, well, it was even more surreal being there, believe me," Gordon said into the mobile telephone that was given to him at Warsaw Chopin International Airport.

"How's the woman?"

"She's fine. Her name's Anna Czolski." When Gordon did not hear anything further, he said, "Hello? Are you still there?"

"Yes, I'm here. You said Czolski?"

"Yeah. She's his daughter. I haven't been able to get much more out of her than that. She said that her instructions from her father were to not say anything until she met with someone named *Newbridge*. I'm assuming that's you."

"Yes, that's right. Listen, how is security?"

"Strong. Two escorts...one Polish, the other a female Brit."

"Good. I'll see you at Dulles."

"Right." Gordon was about to hang up when he said, "Hey, John."

"Yes? I'm here."

"She's beautiful."

There was more silence, and Gordon waited for Angstrom's response. Finally he heard: "I know."

After the call was disconnected, the group walked to a police hub within the airport. The female British agent went into a small, unmarked room, a private bathroom, and ensured that it was clear, and then she came back out and handed Anna a bag. Anna was not sure what to do, or what the bag was for, and as the British agent could not speak Russian, she asked Gordon to translate.

"It's a change of clothes," he told Anna. "She's cleared the room. Go ahead, it's safe."

When their plane landed in Dulles, the standard announcement was made: "Ladies and Gentlemen, for your safety, please remain seated with your seatbelts fastened as the captain pulls the plane into our designated gate."

As the plane slowly taxied along the tarmac, Gordon saw one of the stewardesses approach him. She bent over to whisper, "Sir, my instructions are that you're supposed to come with me."

Gordon nodded to Anna, who was sitting next to him, and she unfastened her seatbelt as well. When they got up to follow the stewardess, their two escorts from Warsaw Chopin followed as well.

The Stewardess turned to them in confusion, extended her

open palm to them, and said, "I'm sorry, it's only this man."

"No, they're all with me," Gordon said. "We need to stay together."

The other passengers watched in confusion, some even in impatient anger, as the group made its way to the front of the plane. When the plane was docked and its door opened, two United States agents were there waiting for them. The captain tried to quickly unfasten his seatbelt (as his co-pilot continued with the docking procedures) in order to get a quick, curious glimpse of his special passengers.

"Are one of these men Newbridge?" Anna asked Gordon as they were escorted through the airport by the two additional U.S. agents that met them at the plane's door.

"No...he decided not to meet us here. You'll see him tomorrow at a safe house."

They walked in silence the rest of the way through the airport, and other than to respond to simple, practical questions such as what she wanted for dinner, there was nothing else discussed with her for the rest of the evening.

A selection of clothes was waiting for her at the safe house, based upon the measurements she had given them in Warsaw, and they were much more stylish than what she had been given previously to change into. When Gordon entered the safe house that evening in order to personally deliver her dinner, he could not believe how stunning she looked in the new clothes.

She knew what he was thinking; she was used to such looks. When his attempts to strike a casual conversation were unsuccessful, he respected her privacy and let her dine alone, and he spent the rest of the evening in the second bedroom of the place watching television.

* * *

The next morning, when Gordon heard a knock at the front door and opened it, he saw a guard standing there who was assigned to the shift, as well as Angstrom, who had just arrived.

"Everything all right?" Angstrom said.

"Yeah, yeah, come on in, John."

They shook hands, with serious looks on their faces in reflection of what Gordon had just been through. Right in the middle of the handshake, they both got big smiles on their faces, and then gave each other a hug and patted each other on the back.

"It's really good to see you again," Gordon said.

"Yeah, same here. Especially in light of what I saw happen on that train."

Gordon nodded his head in earnest. He wanted to say more, but he could tell that Angstrom was anxious to see Anna. "She slept late, and is just now finishing up her breakfast."

"Is she dressed?" Angstrom said as he stepped inside while Gordon nodded to the agent outside and closed the door.

"Yeah. She was up early."

Gordon escorted Angstrom through the apartment to the dining room, and Angstrom could see her from a distance sitting at the table. He was nervous to meet her. When they reached the room, she looked up from her meal and over at them.

Gordon said, in Russian, "Anna, this is John Angstrom... *Newbridge*."

She started to get up, and Angstrom held up his hand and said, "Please, don't get up." Gordon translated for him.

But she did anyway, and she went to him and offered to shake hands as she bowed her head slowly.

When Angstrom took her hand, he could feel her soft skin, but the memory of her covered in blood came to mind, and he said, "I'm sorry for your father."

She turned away for a moment, and then sat down in order to suppress her emotions. When she was able to, she looked up to him and said (this time in slow English), "Thank you. He knew... we knew...that he would not make it."

No one said anything after that, out of respect for her loss.

Finally, she said, "Please, sit down." She motioned to a chair at the table.

Angstrom looked at Gordon, who in response said, "So, she speaks English. I guess you don't need me anymore...I'll leave you two alone." He looked over at Anna, bowed his head slightly, and said, "Anna."

"*Spasibo,*" she responded with a bow of her head.

"You're welcome. It was my pleasure," he said in English, and then with a nod to Angstrom, he left.

Angstrom and Anna looked at each other across the table for a moment. There was so much that he needed to ask her, and then so much more that he did not need to, but wanted to.

She smiled at him, like she knew what he was thinking, and then she said in English, as she reached for the coffeepot on the table, "Let me pour you some coffee."

It was a nice voice, he thought. One that he could listen to for a long time.

* * *

A few days later, *The St. Petersburg Times* reported that Professor Romanov Czolski was found dead in his home, having suffered a massive heart attack. Dmitri read the announcement with interest. Not too much more was reported about the Professor, except that he had no surviving family. There was barely a mention of his academic career, and no mention at all of the many contributions he had made in the field of military technology.

Dmitri was alone, smoking a cigar, drinking cognac, and enjoying the warmth of the Professor's old fireplace. The KGB had ownership to the Professor's home transferred to Dmitri; it was part of his promised reward. He felt extremely fortunate, especially given the tragic end that befell three KGB agents.

But it was not all bad for the KGB. As far as it was concerned, a troublesome Professor was finally silenced, and a nuisance of an agent was no longer a concern. With nothing else to do with the

Professor's house, handing it over to Dmitri was a simple decision. Dmitri was grateful to the KGB, both for the home, as well as for facilitating his promotion to the Professor's position at the University.

Bartok's *The Miraculous Mandarin* played softly in the background, and even at such a low volume he felt the power of the music. It was at the part where the Mandarin first appears in the story, and Dmitri turned the volume higher, because that segment of the piece started *Pianissimo*. The music gave him chills that evening, especially when he recalled the story of the Mandarin and what happened to him.

There was an unsettled feeling in Dmitri. It started ever since he had been manhandled and tied up like an animal. In addition, the Professor's death was beginning to weigh on his conscience. He felt another chill and threw a log into the fire. This was the outcome that he wanted, that he had bargained for, he thought to himself, but now he felt alone, and isolated.

He thought it would feel different after he assumed the Professor's position, like he would be on top of the world, and that it would have a profound effect on him. Instead, he felt empty...hollow...even evil. It didn't help that his fellow professors looked at him begrudgingly. He could feel their stares behind his back, like they were talking about him. As a full Professor, his salary did not change much either, contrary to what Raskalnikov had promised him. Raskalnikov had sold him on a good story, but now the reality had set in: he had been complicit in a heinous series of events. He didn't even know what really happened to the Professor. It was no matter, he tried to convince himself, for at least he was a full Professor now, and he did, after all, own the Professor's house and land.

His eyes moved about the dark room, lit only by the light emanating from the fire, and the stirring music that was playing intensified the strange feeling that he was experiencing. He started to wonder if he was being watched.

He threw his cigar into the fireplace and then went around the room frantically looking at everything and trying to figure out where there might be a hidden camera or microphone. The music was building to a crescendo; it got louder and louder, and then it was *Fortissimo*; the volume was so loud that it disturbed his sensibilities—it began to feel like noise to him—very distressing noise.

The needle of the old record player hit a scratch and began to repeatedly skip at a most frenzied moment in the piece. He moved quickly to the player, removed the needle, picked up the record, and slammed it against the edge of the table, shattering it into pieces. His hands ran through his thick hair as he tried to calm himself, and then he rushed back to the small table by the fireplace, drank the rest of the cognac, and quickly poured himself another.

The room was quiet—nothing but the sound of the crackling fire—and he leaned forward in his chair and put his head in his hands; he was emotionally distraught. He was a troubled soul before the Professor's murder, and he was even more so now.

All of a sudden, the doorbell rang, startling him. It was a man that Dmitri had never seen before. He said nothing when Dmitri opened the door.

"Yes, may I help you?" Dmitri finally said. He did not have a good feeling about seeing the man.

"Hello Dmitri, how are you?"

"I'm sorry, have we met?" Dmitri said, in a state of confusion.

"Such manners—I'm glad. That will make our...working together... more pleasant."

"What...what do you mean? I don't know you."

"No, you don't. But you will. I'm Raskalnikov's replacement, and we've got a lot of catching up to do before we get back to work. May I come in?"

21

Resolution

Angstrom was in a borrowed office at the Pentagon. He had spent most of the day with Anna, and had, in fact, spent the last several days with her, standing witness during interviews and debriefings. He would be going back to her later in the day, but he was told that Garrett needed to speak with him, and that he would be calling, so there he waited.

At almost the exact time that he was expecting the call, Garrett entered the office.

"So, will your young friend be joining you?"

Angstrom was startled by the voice and looked up to see him standing there. He had been through so much over the last several days, and Anna occupied so much of his thoughts, that he had to struggle for a moment to realize what Garrett was talking about. When it finally dawned on him, he thought it was classic Garrett: the mission was over, Gordon had already reported the specifics to Franklin, who in turn relayed them to Garrett, and instead of Garrett saying anything about the mission, he was on to the next pressing issue.

It was all just a test anyway, Angstrom thought to himself, because as far as Garrett was concerned, the *only* issues that really mattered were whether Angstrom would accept the position as handler, and whether he would select Keplar as his new protégé. The way Garrett framed his question meant that he already believed that he knew where things stood as far as the first issue.

"I don't think so," Angstrom finally responded. "I don't think he's workable; he... has a weakness."

Garrett shut the door behind him and sat down. "I know. So why haven't you definitely decided? What's holding up your decision? You think you can rehabilitate him...help him be able to control his pecker?"

There was a long pause between them; Angstrom could see Garrett's chest move up and down as he breathed. He looked older to Angstrom than he did when they had met at the park. He could see the tiredness in his eyes.

"I haven't decided yet...with certainty."

Another long pause, as if Garrett was giving him the opportunity to make up his mind on the spot, right then and there, if he chose to do so. It really didn't matter to Garrett either way, because he would be gone himself soon enough, and it would be Angstrom's dilemma to wrestle with, but he wanted to know. Perhaps he wanted to compare Angstrom's final decision with what he himself would have done if it was his decision to make.

"Hmm," said Garrett. He reached over and grabbed a paperweight that was sitting on the desk. It was Garrett's office that Angstrom was borrowing, and Garrett's desk that he was sitting at, and therefore Garrett's paperweight.

"I think I have another option," Angstrom finally said.

Garrett stopped twirling the paperweight and looked up at him. "Gordon?"

"Yes." Angstrom was a little surprised that Garrett had guessed so quickly. "He's not going to work for Franklin."

"Did he say so?"

"He didn't have to. I know Gordon."

Garrett thought about it, and it made perfect sense to him. After all, Gordon and Angstrom had worked together on many missions over the years. And Angstrom had at least five years on Gordon. Besides, he knew that it was true that a man of Gordon's caliber would never be able to work for a man like Franklin.

"I like it," he finally responded as he looked down and studied the paperweight in his hands. "But you haven't talked to him about it?"

"Not yet. Our communication has been pretty limited up until this point. But he dropped me a couple of hints. I think he'd jump at the chance."

"More like jump *from* Franklin."

There was no smirk or grin on Garrett's face. There rarely was. Angstrom wondered what a man like Garrett would do for relaxation and enjoyment after he retired.

"So, my assumption was correct; you're coming back," Garrett continued when Angstrom did not say anything.

"Yes. I am. I can't stay at the TTO.

It was there—Angstrom could swear he saw it: a slight grin on Garret's face.

"We knew you wouldn't."

Garrett put the paperweight back onto the desk, reached across to offer his hand, and said, "Good luck; you're made for it."

Angstrom nodded and reached over to shake his hand. "Are you done?" he asked.

The handshake was an extended one, and still during it, Garret responded, "Yes. You won't see me again."

Their eyes met, as if to express things that words could not. For two men that had spent so much time together but rarely spoke at great length, except for when it had to do with the cold, objective facts of a mission, it would have been out of character, and a disservice to their relationship, for them to say more. Garrett got up and left without turning back, forever.

* * *

The PrC submission was dropped onto Seavers' desk, startling him. He had not heard Angstrom enter his office.

"Sorry, Bill, I didn't mean to surprise you."

Seavers looked up at the man standing in front of his desk and studied his face: it was changed; leaner; the slight puffiness that had been there was gone; a new intensity was in the eyes.

"There's your winner," Angstrom said, pointing down at the submission sitting on the desk.

Seavers looked down at the thick set of papers bound together, cleanly printed, fresh, without a mark on it. A clear, plastic cover protected it, and Seavers stared at the cover page for a moment before picking it up and flipping slowly through its pages, stopping occasionally to study certain things that jumped out at him. Angstrom stood there patiently in front of the desk and gave Seavers time to look at it. Seavers became so intrigued with what he was studying that he forgot that Angstrom was there, and Angstrom sat down as Seavers continued his review. That lasted for close to ten minutes.

The reference to PrC, and the quote at the end of the submission, had been removed, and it finally occurred to Seavers that there was no identity of a submitter. Upon that realization, he was brought back to the moment and realized that he had been studying the submission for quite some time while Angstrom sat there.

He smiled sheepishly at Angstrom. "Sorry. I suppose I lost myself."

"It's alright. It's understandable in light of the circumstances."

Seavers nodded, and when he saw that Angstrom was not going to say anything further, he put both of his elbows on his desk and clasped his hands together, rubbing them nervously. "So, I suppose this was what everything was about?" One of his eyebrows curled upward, as if to underscore the question and prompt a response.

Angstrom did not take him up on the invitation, and instead said, "Fred has studied the document as well. He'll be able to give you a full briefing."

"All right," said Seavers as he leaned back in his chair. The curiosity to know more about what Angstrom had been doing, and the origin of the new submission placed before him, was at the forefront of his mind, but was tempered by years of experience, enough to know that it was not within his *purview*, and so the boundary that Angstrom had set between them was apparent, and respected. The ability to observe, however, subtle facial expressions and slight body gestures, which he had developed through years of dealing with people of different backgrounds and levels, made Seavers aware of the fact that Angstrom wanted to say more, or to talk about something else, and he waited.

Angstrom was, in fact, struggling with whether he wanted to broach one final topic, and he found himself straddling the division between his temporary layover in the TTO, and the return to his old organization, the CIA. Bridging the two was the PrC submission, which made it difficult for him to forget the prior, and transition, with ties completely severed, to the latter. He was considering whether to raise the topic of a certain female consultant who had been retained by the TTO for quite some time, and the unorthodox influence that she wielded within it. Eventually, he decided not to, the decision being made right at that instant. The redemption of her soul, the purging of her demons, or whatever the case may be, was a matter for which, if it was achievable, would need to be attended to by someone else. Her personal situation was surely complicated, and the surrounding circumstances were politically delicate to say the least, and thus, it was suited for someone with a more direct and ongoing connection with her, or, at a minimum, with the general matter. It was not his world anymore, and was therefore not within *his* purview to make such a judgment or intercession. Besides, he knew Seavers well enough by then to know that Seavers probably had a clue as to what was going on, and if he chose not to do something about it, he probably had his reasons.

"Bill, thank you for the opportunity to work here for the last two years. Getting to work with you and know you was the best part of the job."

"You're welcome, John. And I'm… glad I got to know you…at least to some extent." A slight smile. "It looks like you've turned things around. I suppose I won't be seeing you anymore?"

"I don't know, maybe. Certainly not in my previous capacity here at the TTO; but you never know. The things I learned around here were a lot more in depth than the tech briefings I used to get. Maybe I'll come by and pick your brain every once in a while."

"Somehow I seem to doubt that," Seavers said with another smile, this time more pronounced than the last. "Besides, don't be surprised if I'm gone from here myself before too long."

"Hmm. I see." Angstrom was quiet for a moment, and then he asked, "What's your passion?"

"My wife and I have family in Wyoming."

They talked about that for a while—small talk, just to enjoy each other's company a little longer.

Afterwards, there was really nothing more to be said between them, and they shook hands one last time.

* * *

He picked her up from the safe house himself. He wanted to be the one to escort her to the debriefing again, and to be present while it occurred. The things being learned about her so far were incredible—her background was extraordinary.

They said very little between them as they walked to his car. She wasn't sure yet what to think of him, but she was beginning to form an opinion. He looked into her eyes for a brief moment as they sat in his automobile, and then he put the car into 'Drive.'

"I need to stop somewhere first," he said.

She turned to look at him, but he kept his eyes on the road. She became just a bit nervous and did not know what to expect. Double-crosses and back-stabbing were par-the-course from where she came; she had experienced a demonstration of that at its highest form in all that she had just been through. She looked down at her lap and thought about the situation for a minute, and then she looked over at the face of the man that was driving the automobile. She was done running. She had already made it to the place where she was running *to*. Come what may, she was where she was intended to be. Her father had promised her that she would be cared for...protected. Besides, the opinion she was forming about Angstrom was a good one. She was a good judge of character, and she could see in his eyes that there was something special about him. And she liked his face.

Out of curiosity, and to distract herself, she opened up his glove compartment to see what was in it. To her surprise, she found a collection of old CDs. She took them out and began looking through them, and Angstrom periodically turned to look at her while she did so. She paused when she saw a CD of Tchaikovsky's *Symphony Number Five*, and she looked at it for a long time, just staring at it. Angstrom thought that she was going to cry.

But then she moved past it. Cycling through the rest of the collection, she selected something quite different. Holding it up, she asked, "May we listen?" It was John Coltrane's *A Love Supreme*.

"Absolutely. An excellent choice."

He put the CD into the player for her, and they rode in the car for a long time without speaking, just listening to the mystical sounds of the music, with warm air coming in on them through the automobile's opened windows. It was almost the end of spring and a perfect temperature outside

As if on cue, as soon as the music was over, Angstrom had reached his intended destination. The car slowed, pulled over to the curb of the tree-lined street, and then slowly crept forward for about thirty yards. Anna watched Angstrom as his head turned in all directions, like he was looking for something, or rather, to see if anyone might be watching him; she could tell that something was not right. When the car stopped and the engine was turned off, they waited. She felt compelled to remain silent. Twenty minutes passed, and then Anna saw Angstrom stir. A man had appeared far off in the distance, walking toward them on the sidewalk, and Angstrom pulled out a pair of binoculars to look. His source, the one he had instructed to follow the man in the first place over the course of the last couple of weeks, had told him to expect to see the target at about that time of the day.

He watched as the man walked for some time, and then, about fifty yards in front of the car, he stopped—it was right at the entryway to one of the executive suites. The man looked sideways in each direction, as if he himself was nervous about what he was doing, and then he walked up the entryway to the front door of the suite. It was the same place that Angstrom had previously visited.

After Angstrom saw the man knock, the door opened, and Angstrom could see her through the binoculars: the brown hair up in a bun, red-rimmed glasses, and the deep, red lips. She was wearing a grey overcoat, with nothing on underneath as far as he could tell. Her bare, shapely legs, and the red pumps—even from fifty yards away he recognized her. Then he saw her raise her arm and hold a black, leather horse whip up to Keplar's chin, and slowly cause it to tilt upward. She held it there for a moment longer before grabbing him by the shirt with her other hand and pulling him inside, and then, the door closed behind them.

Anna watched too, as much as she could make out from so far away, and when the door closed, she turned to look at Angstrom, who by that time had brought his hands down to let the binoculars rest in his lap. He looked at her for a brief moment, tossed the binoculars onto the back seat, and then put both of his hands onto the steering wheel. Judging by his brooding reaction, she wondered if perhaps it was his wife who they had just seen answer the door—or maybe his lover.

I haven't decided yet was what he had told Garrett. Well, now he had.

He started the car's engine but did not begin driving right away. As they sat there in the idling car, she decided to ask him. What did she have to lose? She wanted to know. She wanted to learn about him—she was intrigued. In English, modified as it was by her Russian accent, she asked, "Who were they?"

But he did not respond. The car engine continued running as they remained there, parked. Her question, though, did have an effect on him. Angstrom was done with his thoughts on Keplar, and he turned his attention back to the immediate—to Anna.

She's beautiful, he thought. There was no doubt about it in his mind. He turned and looked at her like he did that first day, when he studied the image of her on the CD—when he had first become enchanted by her beauty. For some reason he thought about Garrett, and wondered if *he* was ever married.

Angstrom had become infatuated with Anna when he first saw her image, in awe of a woman that he did not know, and knew nothing about; a woman who, for all intents and purposes, may not have even been real. But now he knew that she was; she was right there with him. She breathed, her skin was pigmented, her body took up space, and she moved; she was absolute. It was surreal to see her, to watch her, and to be with her, to be convinced of her existence. He wanted to reach out and touch her, to re-validate again that Candy Mav, no, that Anna, was real, and there. Her beauty, as he had perceived it in the beginning, preceded her, and now the actual did not fall short of the imagined, the dreamed of. He wanted to open himself up to her, to reveal himself—all of him—his innermost thoughts and feelings.

The more he had communicated with her over the last several days, the more he had become convinced of one of two possibilities: she was perfect and heavenly, in all respects and regards—or, as may have been more likely the case—she was the before, the during, and the after…for him.

And for her, he had been just as mysterious. Alluded to by her father as one who must surely have been unique and significant—

Angstrom had found the impossible, and recognized it for what it was—she had formed some facsimile of him in her mind beforehand. He was the unknown, the metaphysical, and her deliverance, now there with her. Newbridge. For her as well, there was no disproportion between the theoretical, the contemplated, and the actual.

She met his gaze, and at that moment, without any words spoken between them, their two souls truly met for the first time, and were one. The smiles on both of their faces disappeared and were replaced by looks of genuine earnestness, and as she stared into his eyes, she found warmth.

Then, after a meaningful moment of time had passed, a corner of his mouth turned up in a playful smile, and he said, "We're in no hurry today. Let's go somewhere else and have some coffee... and talk. Just you and me, without the others."

She looked at him with a nervous expression on her face, and then with an intensity, until finally a soft smile appeared, and she nodded her head to him and said, "Yes."

ACKNOWLEDGMENTS

A special thanks to my good friend John Ortiz for introducing me to the fascinating field of steganography.

ABOUT THE AUTHOR

Paul J. Bartusiak was born and grew up on the South Side of Chicago. He obtained his BSEE from Tennessee Technological University, his MSEE from the University of Texas at Arlington, and his Juris Doctor from Chicago-Kent College of Law. Over the past 25 years he has worked at Fortune 500 companies, first as an Electrical Engineer, and then as a Corporate Attorney. He lives with his family in Lake Forest, IL. This is his second novel.

Made in the USA
Charleston, SC
10 July 2013